BIOSEQUESTRATION
AND ECOLOGICAL
DIVERSITY

Mitigating and Adapting to Climate Change and Environmental Degradation

Social-Environmental Sustainability Series

Series Editor
Chris Maser

BIOSEQUESTRATION
AND ECOLOGICAL
DIVERSITY

Mitigating and Adapting to Climate Change and Environmental Degradation

Wayne A. White

CRC Press
Taylor & Francis Group
Boca Raton London New York

CRC Press is an imprint of the
Taylor & Francis Group, an **informa** business

CRC Press
Taylor & Francis Group
6000 Broken Sound Parkway NW, Suite 300
Boca Raton, FL 33487-2742

First issued in paperback 2019

ISBN-13: 978-1-4398-5363-4 (hbk)
ISBN-13: 978-0-367-86604-4 (pbk)

Library of Congress Cataloging-in-Publication Data

White, Wayne A.
    Biosequestration and ecological diversity : mitigating and adapting to climate change and environmental degradation / Wayne A. White.
        p. cm. -- (Social environmental sustainability series)
    Includes bibliographical references and index.
    ISBN 978-1-4398-5363-4 (hardback)
    1. Carbon sequestration. 2. Carbon cycle (Biogeochemistry) 3. Acclimatization. 4. Climatic changes--Environmental aspects. 5. Biodiversity conservation. I. Title.

SD387.C37W49 2012
577'.144--dc23   2012025564

This book is dedicated to my brother Kenneth V. White.

His courage and optimism in the most difficult of

circumstances is a source of lasting inspiration.

Contents

Series Editor's Note

In the introduction to this book, Wayne White relates the story of a banker in Kansas who, "as a father and grandfather," was seriously concerned about the available supply of water and the concomitant economy thirty years into the future, but, as an officer of the bank, "he had a fiduciary responsibility to maximize revenue in the current and next fiscal years." He saw his job constrained by the aphorism, "success in business comes from working on the things that you can control." Left unsaid in this statement is the practical fact that our social life is under the relentless tutelage of an economic system, the premise of which is dissatisfaction with one's present standard of living, supported by an army of advertisers who espouse the constant need for ever-greater acquisition—for more, *always more*.

Consequently, our economic system is based on symptomatic thinking focused solely on quick fixes to anything threatening economic expansion and the linear growth of profits. This pattern of thinking, left unchecked, is rapidly altering the global ecosystem in a way that makes much of it progressively less conducive to human habitation—a situation compounded not only by a bourgeoning human population, but also by the increased longevity of human life.

Moreover, we must think systemically about the three interactive spheres that comprise our world: the atmosphere (air), the lithohydrosphere (the rock that constitutes the restless continents and the water that surrounds them), and the biosphere (the life forms that exist within the other two spheres). Although socioenvironmental sustainability requires systemic thinking, we arbitrarily delineate our seamless world into discrete ecosystems nonetheless, as we try to understand the fluid interactions between the nonliving and living components of the earth. But, if you picture the interconnectivity of the three spheres as being analogous to the motion of a filled waterbed, you will see how patently impossible such divisions are because you cannot touch any part of a waterbed without affecting the whole of it.

Consequently, there are *no problems* "out there." The environment is simply a biophysical mirror reflecting our chronic, socioeconomic dysfunction, which is now reaching a critical stage with increasing portents of terminal conditions for supporting human life in the decades and centuries to come in more and larger portions of the global ecosystem.

Is this outcome inevitable? Is there nothing we can do to avert it?

No, it is not inevitable, and yes, we can avert it, but we must act *now*—decisively and collectively. I emphasize *now* because this eternal moment— the here and now— is all we have or *ever will have*. As Wayne illustrates in his book, "there are many ways to reduce greenhouse gas emissions and some fascinating ways to remove them from the atmosphere by changing the way we use and manage land."

To accomplish this, however, we must ground our culturally designed landscapes and seascapes within nature's evolved patterns and take advantage of them if we are to have a chance of creating a quality environment that is both pleasing to our cultural senses and biophysically sustainable. To wit, we must do three primary things: (1) control our human population, (2) refocus our concept of development from the exploitive, symptomatic subjugation of nature to a harmonious, systemic cultural evolution of sustainability with nature, and (3) protect existing biodiversity—including habitats and biophysical processes—at any price for the long-term sustainability of the ecological wholeness and the biological richness of the patterns we create across our global landscapes and seascapes.

For a nation and world that is increasing governed by symptomatic thinking, despite the world's systemic nature, Wayne's book is an important voice in support of a rising global consciousness with respect to protecting and nurturing nature's free services for all generations—rather than simply exploiting them for immediate, personal gain. To this end, he explains how greenhouse gases can be withdrawn from the atmosphere and isolated over time in both land and oceans through a process known as "biosequestration." This process can be thought of as long-term storage accomplished by honoring the biophysical principles through which nature operates. Wayne's message is clear and gentle, yet profound in its simplicity: by working in harmony with nature, we not only begin to heal the damage we humans have caused by degrading the global ecosystem over the centuries, but we also begin to heal ourselves by accepting our role as trustees of the earth as a biological living trust—of which *all* generations are the beneficiaries.

This book makes an important contribution to the series on socioenvironmental sustainability by offering an excellent blueprint of the systemic components and actions necessary to begin a conscious, collective process of healing our home planet, each within our own capacity. The choice of accepting and acting on the wisdom contained within these covers belongs to us, the adults of the world. The consequences of our decisions, however, we bequeath to *all* generations—for better or worse. How shall we choose?

Chris Maser
Series Editor

About the Author

Born in Minneapolis, Minnesota, and raised primarily in the Kansas City metropolitan area, **Wayne A. White** has lived on small farms most of his adult life. He lives with his wife Sandra on an 80-acre farm in Jefferson County, Kansas, where they raise grass-fed beef, apples, pears, berries, biomass energy, and a variety of vegetables and herbs. Forestland health, high-diversity native grassland mixtures, and land management practices that protect and enhance ecological diversity are central interests. White has a PhD in sociology from Kansas State University, has taught sociology and political science, and has worked as a legislative lobbyist, grant writer, and program administrator for a statewide nonprofit legal services organization. He owns and cares for forest and grassland in Kansas, Michigan, and Ontario, Canada.

Acknowledgments

Wayne C. Rohrer, my dissertation advisor and mentor in the sociology department at Kansas State University, told me many years ago that there may be no more sane behavior than planting trees. Dr. Rohrer has been deceased since 1993, but his wisdom and passion for ecological sanity and social equity become more salient as time passes. I am indebted to him for his early recognition of the emergence of ecological scarcity and its many implications for industrial societies.

Robert J. Antonio read the manuscript as it was nearing completion and demonstrated an uncanny ability to identify the issues about which I had doubts and that needed clarification. Our many conversations about climate change and his work on globalization helped to shape this book. Eric Hanley, Mike Lacy, and Frank Svoboda reviewed an early proposal for the book and offered many helpful suggestions. Conversations with each of them are always stimulating and often lead to the reconsideration of ideas and the discovery of new evidence. Ed Reznicek, Frank Svoboda and Carl Gillies provided interviews and time in the field sharing their passion and knowledge of the work they love. John Harms is a friend and colleague who is always supportive and willing to exchange ideas and arguments on any topic.

Joane Nagel's assistance with obtaining full access to the University of Kansas Libraries and the appointment to the Institute for Policy and Social Research was indispensable and much appreciated. Kim Glover at Anschutz library at K.U. showed great patience in teaching me how to fully utilize Endnote for references and citations. Marilyn Rohrer made sense of and provided needed order to long interview transcripts.

The Canadian travel crew of Craig Willis, John Baker, Mike Hanson and Rick Stein listened to many informal monologues on topics included in the book. They share my love and appreciation for the north woods and lakes and help me take breaks from work to enjoy the beauty of nature and friendship. Mike Hanson also takes excellent photographs and provided two of the several found in Chapter 5.

Chris Maser, series editor, helped make this book possible by asking me to write a proposal elaborating the ideas and perspective he heard me speak of at a conference on sustainable actions at Missouri State University. His invitation and encouragement served as a catalyst to focus and organize ideas and information that may have otherwise remained in a much less organized state. The staff at Taylor and Francis has been helpful, professional, and responsive at every step in the process.

Introduction

During the late 1970s, I worked on a research project examining groundwater use in western Kansas. For two years, I interviewed farmers, bankers, and hydrologists about the factors considered when making decisions about using groundwater for irrigated crops, mostly corn. I learned from the hydrologists about the nature and extent of water resources in the Ogallala aquifer underlying western Kansas and much of the Great Plains.

In 1980, I was asked to make a presentation to regional bankers summarizing current hydrological information relevant to irrigated agriculture. The bankers were interested, of course, because of the large capitol inputs required for irrigation wells, pumps, center pivot systems, and related equipment. Indeed, the rapid expansion of irrigated agriculture during the 1960s and 1970s had been a boon to the regional economy and lending institutions.

The data summarized in my presentation included then current estimates of the groundwater resource, pumping rates, and rates of natural recharge. I characterized the long-term groundwater situation as it had been described to me by the hydrologists—as a mining operation. Pumping rates exceeded recharge rates and irrigated agriculture as then practiced would eventually deplete the aquifer. I suggested that the bankers offer incentives to encourage dryland farming in areas where the aquifer held less water and encouraged them to consider investment in water-conserving irrigation technology.

I concluded my talk with a prediction that within thirty years (by 2010) irrigated agriculture in western Kansas and the economic benefits associated with it would be in decline. Since that time there has been a substantial investment in water-conserving irrigation technology, but irrigation has been curtailed or eliminated in many areas of the Ogallala with low saturated thickness.

When I had finished speaking, the president of the bank hosting the event approached me and told me that "as a father and grandfather" he was very concerned about water resources and the economy thirty years in the future. But "as an officer of this bank" he had a fiduciary responsibility to maximize revenue in the current and next fiscal years. Because of that perceived, and very real, responsibility, he believed that he could not take professional actions in the short term to address his long-term concerns about water resources and the economy dependent on them. He indicated that he had learned that success in business comes from working on the things that you can control.

I relate this story because it illustrates in a very concrete way one of the central features of the dilemma we face in addressing the life-threatening global ecological crises now confronting us. In our daily lives, especially in our institutional and work roles, we are often constrained into thinking about and acting upon short-term issues in our immediate environment. Most of us are constrained by the need to make a living in a market economy. Short-term thinking encouraged and even required in work settings limits our perceived options. This short-term thinking is ubiquitous and represents a major obstacle to sustainable actions and long-range planning.

Our collective failure to live sustainably cannot be ascribed principally to individual apathy, sloth, ignorance, or greed. Institutional mandates for short-term revenue and profit as experienced by the banker in the story must be situated in the context of several seemingly insurmountable features of industrial society in the twenty-first century. Foremost among those features is the near unanimity among conventional economists and policy makers, in industrial and "developing" states alike, that endless economic growth is required to meet the increasing material needs and demands of a growing world population.[*]

This worldwide growth imperative is driven as much or more by growth in per capita consumption of material goods than by population growth.[†] This is particularly true in the rapidly expanding economies of China, India, Indonesia, Brazil, and other developing states where large populations have been rapidly increasing per capita consumption of energy and other resources. Economic growth as currently measured is highly correlated with increased consumption of natural resources, increased production of waste, and ecosystem degradation.[‡] This remains true despite some substantial improvements in resource efficiency (production per unit of material input) in industry and agriculture in recent decades.[§] The world industrial economy, often characterized as moving toward an information- or service-based economy, remains highly dependent on fossil fuels and increased throughput of both renewable and nonrenewable resources.

Increased throughput of carbon-based fossil energy, destruction of about half of the Earth's forests and other land use changes have resulted in ever higher levels of waste in the form of carbon dioxide (CO_2) and other greenhouse gases and a diminished capacity of the planet to absorb and store those wastes. Despite long-standing concerns about the dangers of depleting essential natural resources, it is the capacity of the biosphere to absorb wastes that has emerged as the primary limiting factor of world economic growth.[¶]

---

[*] Herman E. Daly, *Beyond Growth: The Economics of Sustainable Development* (Boston: Beacon Press, 1996).

[†] L. R. Brown, *Plan B: Mobilizing to Save Civilization* (Washington, DC: Screenscope, 2010).

[‡] Daly, *Beyond Growth*.

[§] William McDonough and Michael Braungart, *Cradle to Cradle: Remaking the Way We Make Things*, 1st ed. (New York: North Point Press, 2002).

[¶] Tim F. Flannery, *The Weather Makers: How Man Is Changing the Climate and What It Means for Life on Earth*, 1st American ed. (New York: Atlantic Monthly Press, 2005).

The composition of the Earth's atmosphere has been fundamentally altered by human actions. Recent (2011) atmospheric CO_2 levels of about 392 parts per million are approximately 40 percent higher than the 280 parts per million in 1750 when humans started burning large quantities of coal.[*] It was the realization of this profound, human-induced change that prompted Bill McKibben to proclaim the "end of nature" as early as 1989, by which he meant that no part of nature is untouched by civilization. McKibben also was one of the first to point out that since we collectively have altered nature, we have a collective responsibility to restore it to health.[†]

Even prior to the widespread burning of fossil fuels, deforestation, and agricultural practices (plowing and other forms of soil disturbance) had been reducing the Earth's carbon storage capacity. The massive infusion of greenhouse gas emissions from the burning of fossil fuels combined with the reduced capacity of the land system to absorb carbon due to deforestation and other land use practices created the climate and ecological crises that we face today. Avoiding catastrophic global warming and maintaining the habitability of the earth will require both reducing emissions and increasing carbon storage in the land system. This is not a claim of "radical environmentalists." It is an overriding consensus in the published research of geoscientists.[‡]

The ecological trends that currently threaten our collective human future often seem too large, too distant, and too complex to serve as a guide to specific actions. Human-induced climate change may be the quintessential problem of this type. It affects the entire global ecosphere, but in unique ways in each ecological niche. The climate-related changes occurring in the Arctic are more obvious than the changes occurring in the Great Plains, but if one investigates the rapidly accumulating evidence in any ecological area, climate change is ubiquitous and accelerating.[§]

I have talked to many friends, relatives, and acquaintances who explicitly or implicitly indicate that climate change and related ecological crises (deforestation, dying coral, ocean acidification, rising sea levels, accelerated species extinction rates, water shortages, increasing storm intensity, etc.) are too complicated and too distressing to think about for long. They state or imply, "How can I, as one person, affect the melting ice sheets in Greenland and the Antarctic? What can I do that will meaningfully affect atmospheric greenhouse gas levels and climate change?" This is an understandable response,

---

[*] Gaia Vince, "An epoch debate," *Science* 334, no. 6052 (2011): 32–37.

[†] Bill McKibben, *The End of Nature*, 1st ed. (New York: Random House, 1989).

[‡] Christopher B. Field and M. R. Raupach, *The Global Carbon Cycle: Integrating Humans, Climate, and the Natural World* (Washington, DC: Island Press, 2004); Brian J. McPherson and E. T. Sundquist, *Carbon Sequestration and Its Role in the Global Carbon Cycle* (Washington, DC: American Geophysical Union, 2009).

[§] Alun M. Anderson, *After the Ice: Life, Death, and Geopolitics in the New Arctic*, 1st ed. (New York: Smithsonian Books, 2009); James E. Hansen, *Storms of My Grandchildren: The Truth About the Coming Climate Catastrophe and Our Last Chance to Save Humanity*, 1st U.S. ed. (New York: Bloomsbury USA, 2009).

and one that resonates with many well-intentioned people. I attempt to identify some meaningful actions in later chapters that almost anyone can accomplish.

There are others who deny the reality of greenhouse gas pollution and its effects on climate. I think that their numbers will continue to dwindle as the evidence of anthropogenic climate change accumulates, but some will, no doubt, cling to alternative explanations or stubborn denial for some time. One can hope that the deniers and contrarians will make a contribution to better scientific understanding of climate change. Obviously, we cannot count on them to help in solving a crisis that they either deny exists or insist is not caused by or amenable to human action. Unfortunately, corporate entities that profit from the continuation of fossil fuel–based energy systems and others with strong financial and ideological commitments are willing to spend large amounts of money supporting deniers and publicizing their views.[*]

Many people, perhaps the majority, have chosen, by their actions or inaction, to ignore the problem and stay busy with other more immediate and more manageable issues. This appears to be the predominate approach, especially in the United States, where citizens concerned with public issues have been largely replaced by consumers understandably concerned with their personal and familial economic well-being and survival. It is these consumers who are the primary target audience of the corporate-funded publicity referenced previously. Some have recognized the reality of anthropogenic climate change, but apparently have given up on the possibility of doing anything about it and have resigned themselves to a catastrophic future for their (and our) children and grandchildren.

This book is intended to be useful to those who have not resigned themselves to a bleak ecological future that is entirely beyond our collective ability to affect in a positive way. There are many ways to reduce greenhouse gas emissions and some fascinating ways to remove them from the atmosphere by changing the way we use and manage land. Both of these strategies, implemented in a vast diversity of ways, will be required if we are to avoid dangerous global warming of 2°C or worse. Chapter 1 provides an overview of recent scientific evidence on anthropogenic global warming and associated ecological changes. There is a massive amount of research and evidence on this topic. I summarize it here to provide an update of recent findings and remaining uncertainties and to provide a context for the discussion of the carbon cycle and land management that follows.

My focus throughout the book is on evidence and practices related to the removal or sequestration of greenhouse gases from the atmosphere. More

---

[*] Naomi Oreskes and Erik M. Conway, *Merchants of Doubt: How a Handful of Scientists Obscured the Truth on Issues from Tobacco Smoke to Global Warming*, 1st U.S. ed. (New York: Bloomsbury Press, 2010); James Lawrence Powell, *The Inquisition of Climate Science* (New York: Columbia University Press, 2011).

specifically, this book is about biosequestration, the process of removing CO_2 from the atmosphere and storing it in the land system carbon reservoir through biological processes managed and nurtured by human action. Technically, biosequestration includes storing carbon in both the land and ocean reservoirs through photosynthesis, but human action to enhance carbon storage in the land reservoir has much more potential, at this time, than attempting to affect carbon uptake in the ocean reservoirs. Chapter 2 describes the carbon cycle and how it has been transformed by the burning of fossil fuels, deforestation, and other destructive land use practices.

Chapter 3 describes biosequestration, identifies essential principles of restorative land management practices, and assesses the current magnitude, stability, and potential of carbon storage in the Earth's forests, grasslands, croplands, and wetlands. Chapter 4 provides examples of proven biosequestration strategies that have been implemented in a variety of places and ecosystems. Learning about concrete examples is a good way to get ideas about strategies that you might want to try on your land or in your community. There are several basic principles that apply to biosequestration anywhere in the land reservoir. These principles are illustrated through examples from the African Sahel, an organic farm in Kansas, and from the grasslands and forests that I care for and manage in Kansas, Michigan, and northwest Ontario, Canada.

Changes in individual behavior alone will not be sufficient to make the changes needed in greenhouse gas emissions, ecological restoration, or enhanced biosequestration of carbon in the terrestrial system. We will need changes in energy, agricultural, environmental, and economic policies at local, state, national, and international levels. A full examination of these complex policy areas is outside the scope of this book. Energy system changes and policy are discussed briefly in several chapters as they relate to needed changes in land use and land management practices. The emerging scientific consensus that fossil fuels use must be rapidly reduced and virtually eliminated from the global energy system provides an important context for effective, rapid, and less expensive land management changes.*

Chapter 5 identifies some existing and effective conservation policies that provide technical assistance and financial resources for ecological protection and restoration. The implications that energy system transformation and widespread ecological restoration have for the future of industrial capitalist societies are explored in the context of the dilemma of economic growth. That dilemma involves the contradictory realities of a continued, near-universal pursuit of economic growth, with growth in energy, material, and waste throughput in a world economy that has already exceeded the

---

* McPherson and Sundquist, *Carbon Sequestration*.

ecological carrying capacity of the planet.* The dilemma of growth is made more perplexing by the very real needs of a growing global population that reached 7 billion in 2011 and is expected to reach 9 billion by 2050.†

If you own or manage land there are inexpensive, proven management practices to increase the biosequestration of carbon and ecological diversity in and on your land. The amount of carbon you can sequester depends on how much land you manage, where it is located, what is currently growing on it, how soon you begin to make changes, and the kinds of changes you make. Every piece of land is different and there are numerous ways to approach each place and ecosystem. Despite this diversity, there are clear principles of ecologically sound land management that apply across ecosystems. Even if you own or rent an apartment or city lot, it is possible to make a small but meaningful contribution. If you own or manage a farm, forest, park, grassland, or even a golf course, you can make a larger difference.

Each chapter, especially the last two, includes information and views gleaned from informal conversations and interviews with knowledgeable, passionate persons with practical experience and expertise. I believe that interviews can lend a humane and personal dimension to what might otherwise be dry, impersonal information. The individuals chosen for interviews are passionate about their work and care deeply for the land. This is as true for those engaged in scientific research as it is for those who are activists and farmers. If we are to collectively make positive land management changes to avoid the most catastrophic effects of the climate and ecological crises confronting us, we will need passion and perseverance.

References

Anderson, A. M. 2009. *After the Ice: Life, Death, and Geopolitics in the New Arctic*, 1st ed. New York: Smithsonian Books.

Brown, L. R. 2010. *Plan B: Mobilizing to Save Civilization*. Washington, DC: Screenscope.

Daly, H. E. 1996. *Beyond Growth: The Economics of Sustainable Development*. Boston: Beacon Press.

Field, C. B., and M. R. Raupach. 2004. *The Global Carbon Cycle: Integrating Humans, Climate, and the Natural World*. Washington, DC: Island Press.

Flannery, T. F. 2005. *The Weather Makers: How Man Is Changing the Climate and What It Means for Life on Earth*, 1st U.S. ed. New York: Atlantic Monthly Press.

---

* Donella H. Meadows, Jørgen Randers, and Dennis L. Meadows, *Limits to Growth: The 30-Year Update* (White River Junction, VT: Chelsea Green Publishing, 2004).

† United Nations Food and Agriculture Organization, *The State of the World's Land and Water Resources for Food and Agriculture: Summary Report* (Rome: United Nations Food and Agriculture Organization, 2011).

Hansen, J. E. 2009. *Storms of My Grandchildren: The Truth About the Coming Climate Catastrophe and Our Last Chance to Save Humanity*, 1st U.S. ed. New York: Bloomsbury USA.

McDonough, W., and M. Braungart. 2002. *Cradle to Cradle: Remaking the Way We Make Things*, 1st ed. New York: North Point Press.

McKibben, B. 1989. *The End of Nature*, 1st ed. New York: Random House.

McPherson, B. J., and E. T. Sundquist. 2009. *Carbon Sequestration and Its Role in the Global Carbon Cycle*. Washington, DC: American Geophysical Union.

Meadows, D. H., J. Randers, and D. L. Meadows. 2004. *Limits to Growth: The 30-Year Update*. White River Junction, VT: Chelsea Green Publishing.

Oreskes, N., and E. M. Conway. 2010. *Merchants of Doubt: How a Handful of Scientists Obscured the Truth on Issues from Tobacco Smoke to Global Warming*, 1st U.S. ed. New York: Bloomsbury Press.

Powell, J. L. 2011. *The Inquisition of Climate Science*. New York: Columbia University Press.

United Nations Food and Agriculture Organization. 2011. *The State of the World's Land and Water Resources for Food and Agriculture: Summary Report*. Rome: United Nations.

Vince, G. 2011. An epoch debate. *Science* 334(6052):32–37.

1

Global Warming and Ecological Degradation

1.1 Emergence of the Anthropocene

In 2002 chemist Paul Crutzen argued that we should no longer consider ourselves to be living in the Holocene epoch, the 11,500 year warm and stable period since the last ice age. He used the term Anthropocene, the age of man, to describe a new geologic epoch characterized by human domination of the planet and much less climatic stability.[*] His Nobel Prize–winning work on the chemistry of ozone depletion in the stratosphere showed that humans have demonstrated the capacity to fundamentally and dangerously alter the composition of the Earth's atmosphere.

In 2009 the International Commission on Stratigraphy, the body with responsibility for formally designating geological time periods, determined that Crutzen's Anthropocene concept had merit and began formal deliberations through an Anthropocene Working Group.[†] The group is examining a range of human impacts on the planet and the extent to which human-induced changes in biology, chemistry, and geography will impact the geological record. Some of their observations are sobering. Humans have modified 80% of the Earth's land surface and we now use about 40% of the land surface to produce food. More trees are in plantations or dispersed on farmland than in forests. More than 90% of global vertebrate biomass is now made up of humans and domesticated animals; the total 10,000 years ago was 0.1%. Earth may well be in the early stages of the sixth mass extinction event in its 4.5 billion year history, this one due to human actions. Dams, roads, mines, and cities will certainly be persistent in the geological record. Mines and other excavations move four times more sediment each year than the planet's rivers and remaining glaciers.[‡]

The ozone hole in the upper atmosphere due to chlorofluorocarbon pollution (which persists today without much public recognition) is only one of the long-lasting chemical alterations humans have brought to the Earth's air, water, land, and life. A massive infusion of light carbon, the C-12 isotope

---

[*] Paul J. Crutzen, "Geology of mankind," *Nature* 415, no. 6867 (2002).
[†] Gaia Vince, "An epoch debate," *Science* 334, no. 6052 (2011).
[‡] Ibid.

from burning fossil fuels, has measurably altered the carbon composition of sea shells, coral, and plankton and will be preserved in the fossil record. A range of chemicals that do not exist in nature, like polychlorinated biphenyls (PCBs) and plastics, are now common. Nitrates in agricultural runoff have created dead zones that affect 250,000 square kilometers (96,500 square miles) of the Earth's oceans. The oceans are more acidic than they have been for at least 8,000 years as anthropogenic carbon emissions are dissolved in them. Atmospheric carbon dioxide (CO_2) has increased from about 260–280 parts per million throughout most of the Holocene epoch to 392 parts per million in 2011.[*]

To most contemporary scientists, Crutzen's observations about the centrality of human impacts on nature and the list of human changes to the planet under consideration by the Anthropocene Working Group are apt characterizations of contemporary reality and not particularly controversial. This has not always been the case. The widespread scientific recognition that human actions have profound global effects on nature, and the atmosphere in particular, is relatively recent. A common belief among scientists during the middle years of the twentieth century, especially geologists and meteorologists who studied weather and climate, was that human actions were inconsequential or, if measurable at all, would not have a significant impact on the Earth's climate until some time in the distant future.[†] Burning coal, it was argued, might cause local or even regional air pollution, but it could not alter the composition of the global atmosphere. Deforestation and agricultural practices that alter the landscape might change local weather patterns, but the idea that these human-induced changes could alter global climate was considered far-fetched. CO_2, in particular, was commonly dismissed as a potential factor in warming the planet because it was such a small part of the Earth's atmosphere and reliable measures of increasing atmospheric concentrations of the gas were not available.[‡]

The view of nature as vast, self-regulating, inherently stable, and impervious to human action was so widespread that it was considered common knowledge. In his work on the history of climate science, physicist and historian of science Spencer R. Weart characterized this set of traditional beliefs as so widely accepted that it was considered part of a universal principle: the "balance of nature." The belief that human actions were too weak to upset the balance of nature was widely shared by scientists and the public and was supported by "religious faith in the God-given order of the universe."[§]

The scientific "consensus" in the mid-twentieth century that considered human action in general, and CO_2 emissions from burning fossil fuels in particular, insignificant in relation to global climate change was based on

---

[*] Vince, Gaia, "An epoch debate."
[†] Weart, Spencer R., *The Discovery of Global Warming*, rev. and expanded ed. (Cambridge, MA: Harvard University Press, 2008).
[‡] Weart, *The Discovery of Global Warming*.
[§] Weart, *The Discovery of Global Warming*, 8.

largely untested assumptions and a lack of countervailing evidence rather than a body of research. Research and evidence about climate change were about to enter a new phase of rapid development, initiated in large part by the availability of reliable new data on average global temperatures and increasing concentrations of atmospheric greenhouse gases.

The onset of the Anthropocene is often viewed as coinciding with the onset of the Industrial Revolution in about 1750–1800 AD. It was at this time that humans had begun mining and burning large quantities of coal and thereby interrupting the natural carbon cycle by introducing substantial quantities of methane (CH_4) and CO_2 into the atmosphere. We know also that humans have been engaging in large-scale soil disturbance for at least 10,000 years through agricultural activities. Early wet agriculture, such as flooded rice paddies and irrigated taro practiced in much of Asia, produced large quantities of CH_4. Air bubbles trapped in the Greenland and Antarctic ice sheets 8,000 years ago show clear evidence that human agriculture and land management practices had begun to alter natural CO_2 and CH_4 emissions long before the industrial revolution.[*]

Whether we date the Anthropocene epoch, the age of human domination, from the beginning of agriculture or the onset of the Industrial Revolution, the agricultural and industrial systems created by humanity have transformed the planet and are now unequivocally the most powerful influences on contemporary global warming, climate change, and a full range of related ecological changes. The reader cannot judge the truth of this assertion, or any other about material conditions, without examining the relevant evidence. Human-induced (anthropogenic) climate change has become politicized in the United States to such an extent that the rapidly accumulating evidence on the subject is often ignored. Some of the many reasons for this will be examined as we proceed.

1.1.1 Ecological Constraints

In the Anthropocene epoch humans have unwittingly, in the pursuit of progress, growth, and profit, created the ecological parameters within which we must now live and work. We have no choice about the nature of the circumstances we face. Global warming, climate change, and a range of profound interrelated ecological problems such as ocean acidification, sea level rise, loss of biological diversity, and soil degradation threaten human societies and the survival of many of the species with which we share the planet. Because of the intricate interdependencies of our fragile yet resilient ecosphere, we are as dependent on many of those species as they have become on us. The meaning of Anthropocene is that these circumstances have been set in motion by human actions. Human industry, agriculture, and land management have

---

[*] Tim F. Flannery, *The Weather Makers: How Man Is Changing the Climate and What It Means for Life on Earth*, 1st U.S. ed. (New York: Atlantic Monthly Press, 2005).

fundamentally altered the composition of the Earth's atmosphere and oceans and degraded the ecological systems upon which we depend. Fortunately we have a wide range of choices regarding the actions we can take to mitigate and adapt to the conditions we have created. Unfortunately the effectiveness of those choices and actions in avoiding catastrophic outcomes is an open question, and evidence supporting optimism that increasingly dangerous outcomes can be avoided is scarce.[*]

The changes we have set in motion are happening now, so one choice, at least, is no longer possible. It is not possible to go back to the world that once existed or even to preserve the conditions that exist today. We collectively have set in motion a set of changes to the planet that cannot be undone, even in the time frames studied by geologists. Recognition that humans have the capacity to cause profoundly destructive ecological impacts should not be mistaken for the hubristic notion that humans are now capable of controlling nature in desired directions. I will argue instead that we can begin to slow the degradation that characterizes the Anthropocene if we learn from natural ecosystems, halt destructive practices as much and as soon as possible, and actively work to restore functioning, diverse ecosystems that have been severely compromised by human actions. Controlling nature is not a realistic option.

Some great scholars have offered guidance to those searching for a sound basis for problem selection and a purpose for scholarly work and practical action. The eminent sociologist C. Wright Mills challenged practitioners of his discipline to develop the best possible knowledge relevant to the central problems extant in the historical epoch in which they live and work.[†] I believe that there is sufficient evidence to conclude that anthropogenic warming, climate change, and associated ecological degradation now constitute the central problems of the Anthropocene epoch. The philosopher and social theorist Antonio Gramsci asserted that scholarly work should be directed toward knowledge and an interpretation of the world as a guide to action.[‡]

It is in this spirit that I have undertaken the task of understanding and characterizing the nature of the climate and ecological dilemmas facing us. This chapter and Chapter 2 are intended to describe the origins and some of the paradoxical complexities of "the problem" we have collectively created. The remainder of the book is devoted to identifying some existing and emerging strategies and actions that change the way we use and care for land. Why land? Ecologically sound land management has underappreciated potential to improve our chances of mitigating and successfully adapting to the climate and related changes that are now occurring and likely to intensify for at least several decades.

---

[*] Tim F. Flannery, *Now or Never: Why We Must Act Now to End Climate Change and Create a Sustainable Future*, 1st ed. (New York: Atlantic Monthly Press, 2009).

[†] C. Wright Mills, *Images of Man: The Classic Tradition in Sociological Thinking* (New York: G. Braziller, 1960).

[‡] Antonio Gramsci and Joseph A. Buttigieg, *Prison Notebooks*, Vol. 1, (New York: Columbia University Press, 1992).

1.2 Climate Change as Context

Developing an understanding of the primary factors contributing to anthropogenic climate change and associated ecological decline, continually amended as new evidence and knowledge emerges, will be required if we are to make informed choices about personal and collective actions to mitigate and adapt to our emerging new circumstances. The term global warming describes a persistent increase in the Earth's average temperature in relation to long-term conditions. Most climate scientists use five or ten year temperature averages when discussing and graphically displaying temperature changes or anomalies. This method avoids the distortions (noise) of annual variation and more clearly demonstrates trends over decadal and longer time frames. These changes are often expressed as temperature anomalies in comparison to the average temperature over a specified time period. The warming trend documented over the past thirty years by the National Oceanic and Atmospheric Administration (NOAA), for example, is typically represented as a temperature anomaly in comparison to the average temperature for a specific period, like the twentieth century or 1961 to 1990.

NOAA, the National Aeronautics and Space Administration's Goddard Institute for Space Studies (GISS), the United Nations Environment Programme (UNEP), the Intergovernmental Panel on Climate Change (IPCC), the National Academy of Sciences (NAS), and a number of other national and international scientific organizations have extensive ongoing research programs on climate change. Their research findings are accessible online and are updated regularly. All of these institutions summarize and analyze a rapidly growing body of scientific evidence on global warming and climate change from a wide and varying array of data sources. Each of these organizations and nearly every official scientific body and university with an atmospheric sciences program throughout the world, including the NAS of each industrial nation, have reached the same overriding conclusion: *the Earth's climate is warming and human behavior is very likely the primary cause.* More precisely, the unequivocal rise in the Earth's average temperature since the Industrial Revolution, and especially during the past three decades, "cannot be explained by natural processes alone. Emissions of greenhouse gases from human activities, like burning fossil fuels, must be included to account for the observed warming."*

Figure 1.1 shows in the bar graph (a) that CO_2 from fossil fuels and other sources (primarily concrete manufacturing) is by far the most abundant of the persistent atmospheric greenhouse gases. CO_2 from deforestation and decay of biomass, peat, and other nonfossil carbon releases from terrestrial and marine ecosystems is the second largest portion. CH_4 and nitrous oxide

---

* Alexandra Witze, "The final climate frontiers: scientists aim to improve and localize their predictions," *Science News* 178, no. 12 (2010).

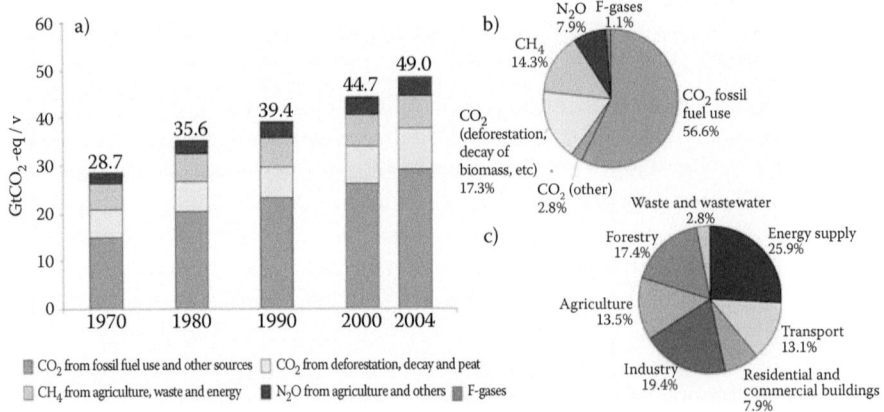

FIGURE 1.1 (See color insert.)
Greenhouse Gases and Sources. *Source*: IPCC (2007).

(N_2O) are the next most prevalent and are shown as carbon dioxide equivalents (CO_2-eq). The column totals are in gigatons (1 billion metric tons; see Appendix A for measurement conversions) of carbon dioxide equivalents.[*]

The upper pie chart (b) illustrates the relative volume of greenhouse gases in more detail. (F-gases include a wide range of trace industrial gases from halocarbons to chlorofluorocarbons.) The lower pie chart (c) identifies the sources of greenhouse gases by industry or human activity. Note that agriculture and forestry combined were identified by the IPCC as the source of 30.9% of all greenhouse gases in these data from 2004. We will return to these terrestrial ecosystems as our central focus.

Since 2004 and the release of the IPCC's Fourth Assessment Report in 2007, global greenhouse gas emissions have continued to increase and a wide range of warming-related changes in the Earth's climate and ecology are now occurring decades to a century earlier than predicted even a few years ago.[†] Some of the climate changes associated with global warming are likely to continue for hundreds, probably thousands, of years and will require substantial adaptation on the part of human societies and many other species. Let me repeat this point because I often hear it said by students and others that climate change is "reversible." The warming and ecological changes caused by anthropogenic greenhouse gases, particularly CO_2, the most persistent of the major greenhouse gases, will continue for decades and probably hundreds, possibly thousands, of years. Sea level rise and ocean acidification, for example, are accelerating and will, by their very nature, continue for centuries, probably longer. This is true even if we were collectively able to

---

[*] IPCC, Climate Change 2007: *Synthesis Report*, Fourth Assessment Report (AR4) (Geneva: Intergovernmental Panel on Climate Change, 2007).
[†] Flannery, *Now or Never*.

somehow entirely stop new CO_2 emissions immediately. Current estimates are that we have thus far experienced only about half of the eventual warming that will be caused by the CO_2 already emitted by human actions. The rate of global greenhouse gas emissions is increasing, however, and will very likely continue to increase for at least the next few years.[*]

The majority of extra heat being added to the climate system is from the buildup of greenhouse gases in the atmosphere emitted by the burning of fossil fuels. The secondary source of anthropogenic greenhouse gases is degradation of the land carbon sink through deforestation and agricultural and land management practices. These actions reduce biomass, release soil organic carbon, and introduce greenhouse gases into the atmosphere in the form of CO_2, CH_4, N_2O, and several others. Although the primary focus of this book is the degradation of the land system and what we can do to reverse or at least minimize future degradation, it is important to recognize that these two causes are intimately related. Both fossil fuel emissions and degradation of the biosphere are central aspects of industrial societal organization, and both must be addressed to mitigate and adapt to global warming and climate change. The majority of scientific, economic, and policy work on climate change is understandably and appropriately focused on ways to mitigate global warming by reducing greenhouse gas emissions caused by the burning of fossil fuels. This is appropriate because fossil fuels are by far the largest source of the increasing concentration of atmospheric CO_2 and several other greenhouse gases.

As we shall see, changes in land management practices are particularly important as ways to adapt to the climate changes that are already having a profound impact on forests and agricultural lands over much of the Earth's surface. Adopting agricultural and forestry practices that enhance the biosequestration of CO_2 and the retention of soil organic carbon and improve ecological diversity can make a meaningful contribution to both adaptation and mitigation. Such practices can reduce the concentration of atmospheric greenhouse gases (mitigation) and create or preserve soils that are more resilient (adaptation) to the increased incidence of both drought and the extreme precipitation events that are now occurring on every continent.[†]

The climate system is imbalanced because the heat entering the troposphere (the approximately 6 miles of atmosphere closest to the Earth) and reaching the surface of the Earth in the form of solar ultraviolet (shortwave) radiation is greater than the heat leaving the troposphere and surface in the form of infrared (longwave) heat. In essence, this is predominately because humans have interrupted the carbon cycle on a huge scale and caused an abnormally rapid increase in the concentration of atmospheric greenhouse gases. The

---

[*] Brian Dawson and Matt Spannagle, *The Complete Guide to Climate Change* (New York: Routledge, 2009).

[†] United Nations Environment Programme, *Climate Change Science Compendium 2009*, ed. Catherine P. McMullen (Nairobi, Kenya: United Nations Environment Programme, 2009).

increasing concentrations of greenhouse gases block increasing portions of longwave infrared heat as it radiates from the surface of the Earth toward the stratosphere (the outer portion of the atmosphere). The result is a cooling of the stratosphere and a warming of both the troposphere and the Earth's surface. There are, of course, a number of other factors and climate feedbacks. I will summarize many of these later, but recent evidence is clear that the overriding source of the climate system imbalance is the concentration of greenhouse gases introduced into the atmosphere by human actions.[*]

1.2.1 Climate Inertia

The majority of the surplus heat caused by the climate system imbalance is going into the oceans. Although most of this heat is accumulating in the oceans' upper layers, warming has been measured as deep as 6,000 feet below the surface. This heating of the oceans is responsible for much of the sea level rise that has been observed. Ocean water expands as it is heated, termed thermal expansion. Melting of land-based ice is responsible for most of the remainder of sea level rise, with a small possible contribution from runoff of irrigation water from aquifers.[†]

Because of their mass, oceans warm and cool slowly. This is a primary reason for climate inertia, the lag in the entire climate system between the time of the introduction of a systemic imbalance and the full realization of its consequences. Inertia due to the large mass of the oceans and the long-lasting nature of CO_2 and some of the other greenhouse gases in the atmosphere are the central reasons that anthropogenic global warming and associated climate changes will continue long—centuries to millennia—after humans cease burning fossil fuels.[‡] One particularly long-lasting impact is the acidification of the oceans due to absorption of CO_2 from the atmosphere. CO_2 is a mildly acidic gas that forms carbonic acid when it dissolves in seawater. The source of most of this excess CO_2 has been tracked with the use of radioactive isotope technology to fossil fuel sources. The ability of the oceans to absorb CO_2, and thus remove it from the atmosphere, declines as the oceans warm and as they become more acidic.[§] Relatively cold and alkaline water can absorb more CO_2 than warmer more acidic water. Acidification damages the ability of mollusks and other sea life to form adequate calcite shells and skeletal material and may have a profound impact on marine food chains. This process is expected to take thousands of years to reverse.

---

[*] Andrew A. Lacis et al., "Atmospheric CO2: principal control knob governing Earth's temperature," Science 330, no. 6002 (2010): 356–359.

[†] Dawson and Spannagle, *The Complete Guide to Climate Change*.

[‡] Committee on Stabilization Targets for Atmospheric Greenhouse Gas Concentrations, "Climate Stabilization Targets: Emissions, Concentrations, and Targets Over Decades to Millenia," http.www.nap.edu/catalog

[§] Christopher L. Sabine et al., "The oceanic sink for anthropogenic CO2," *Science* 305, no. 5682 (2004).

The conservative assessment on the issue of committed warming by the IPCC is that, based on previous emissions, we are fully committed to additional warming until 2030, and about 33% of additional warming until 2050, and 20% of warming until 2100 will be due to greenhouse gases emitted prior to the start of the twenty-first century.[*] The rate of increase or decrease in the global volume of emissions and the total atmospheric concentrations of greenhouse gases will be a primary determinate of the degree of warming over the next century and beyond. Because of the persistent nature of CO_2 in the atmosphere, cumulative emissions are a primary determinant of future warming.

1.3 An Overview of Climate Change Evidence

The National Oceanic and Atmospheric Administration publishes a report each July titled "State of the Climate." The report relies on the research of more than 300 scientists from 160 research groups in 48 countries. The 2009 and 2010 "State of the Climate" reports concluded that the past decade was the warmest in the instrumental record.[†] This finding was not based on climate model projections or estimates. It was derived from measurements and observations of thirty-seven planetwide features. Eleven of the key indicators used by NOAA to document a warming world are shown in Figure 1.2. NOAA utilizes a variety of measures collected from diverse sources, including satellites, weather balloons, weather stations, ships, buoys, and field surveys.

The factors included in Figure 1.2 serve as a good example of the most fundamental indicators of a warming planet. The annual NOAA "State of the Climate" report provides a comprehensive overview of both natural and anthropogenic climate variables with an emphasis on changes over the past thirty–fifty years. All of the indicators in Figure 1.2 are in the direction of a warmer planet. Those who question the scientific evidence and consensus on global warming often limit the debate to disputes about land surface air temperature records, which serves to divert attention from the broad range of measures available.[‡] Land surface air temperature, sea surface temperature, marine air temperature, sea level, tropospheric temperature, ocean heat content, and specific humidity are all increasing as measured by multiple time series datasets. Northern Hemisphere ice cover in March–April, September Arctic sea ice extent, glacier mass balance, and stratospheric temperature all

---

[*] IPCC, *Climate Change 2007*.
[†] J. Blunden, D. S. Arndt, and M. O. Baringer, eds., "State of the climate in 2010," *Bulletin of the American Meteorological Society* 92, no. 6 (2011): S1–S266.
[‡] J. J. Kennedy et al., "How do we know the world has warmed?," in "State of the climate in 2009," *Bulletin of the American Meteorological Society* 91, no. 7 (2010): S26–S27.

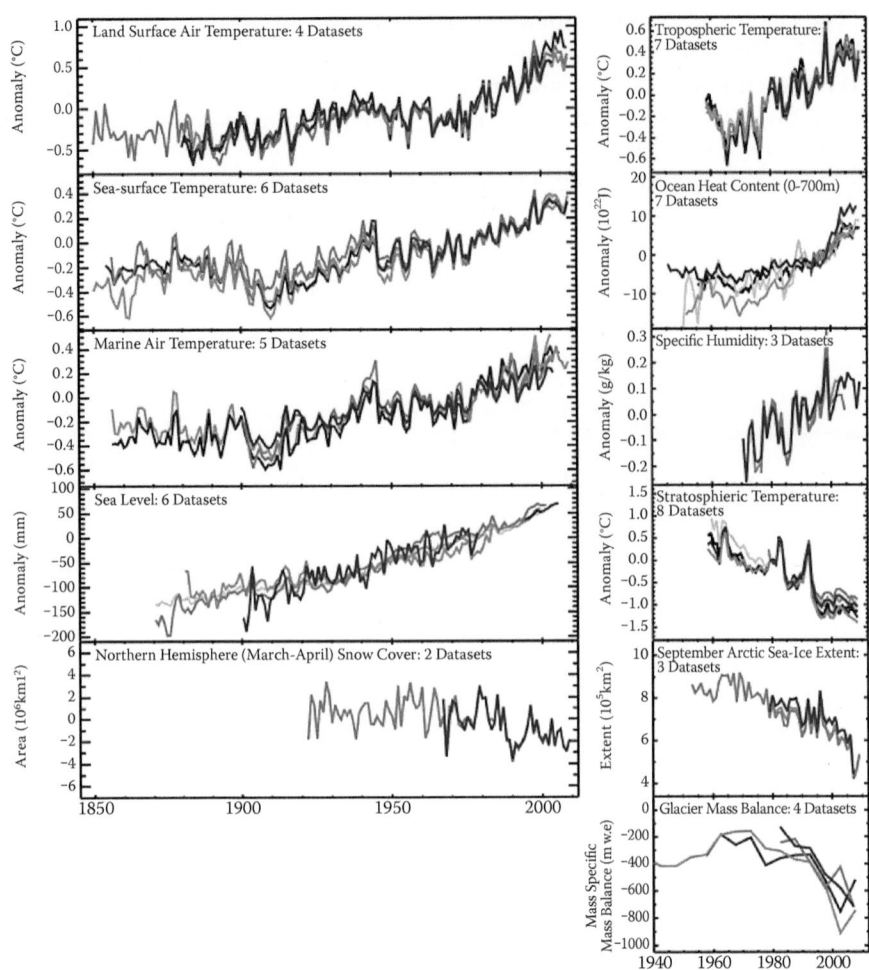

FIGURE 1.2 (See color insert.)
Warming Indicators. *Source*: National Aeronautics and Space Administration (http://www. ncdc.noaa.gov/bams-state-of-the-climate) D. S. Arndt, M. O. Baringer, and M. R. Johnson, eds., "State of the climate in 2009," *Bulletin of the American Meteorological Society* 91, no. 7 (2009): S1–S224.

show a long-term downward trend. Stratospheric cooling, sometimes cited as an inconsistency in climate data, is an expected consequence of greenhouse gas increases.[*]

Temperature data collected by the Goddard Institute for Space Studies (GISS) are also derived from the instrumental record, dating back to 1880. Average global surface temperature measures are obtained by combining

---

[*] Kennedy et al., "How do we know?."

land-based meteorological station data with sea surface temperatures (from ships in the early years and from satellite measurements in recent decades) and those obtained from Antarctic research stations. GISS analysis shows average global surface temperature has increased approximately 0.8°C (1.5°F) from 1880 through 2009.

The majority of measured warming has occurred in the past thirty years. The global average temperature has been rising by about 0.2°C (0.36°F) per decade for the past three decades. The last decade, through 2010, was the warmest in the instrumental record, which goes back to 1880. The last year of the decade, 2010, was tied with 2005 as the warmest year on record, with 1998 a close second according to both GISS and NOAA. Using slightly different measurements, the UK's Meteorological Office and the University of East Anglia rank 2010 as the second hottest year on record behind 2005, with 1998 ranked third. The World Meteorological Organization reviewed all three analyses and noted that 2010, 2005, and 1998 were very close to the same temperature and rank as the hottest years in the instrumental record.[*]

The combined land and surface temperatures across the planet in 2010 were 0.62°C (1.12°F) higher than the average for the twentieth century. In the contiguous United States, average temperatures were 0.6°C above the twentieth-century average in 2010. It was the warmest year on record in the Northern Hemisphere and the sixth warmest in the Southern Hemisphere. The GISS found that the global average temperature in 2010 was 0.74°C (1.33°F) warmer than the average for the period from 1951 to 1980. Despite slight variations, the warming trend has been remarkably consistent in recent decades, as 2010 was the thirty-fourth consecutive year in which global average temperatures were warmer than the twentieth-century average.[†]

Except for a leveling off from the 1940s through the 1970s, the warming trend since 1880 has been fairly consistent. The apparent reason for this thirty-year hiatus in the long-term warming trend serves as an instructive introduction to complications introduced by atmospheric aerosols (solid or liquid particles). The period from the 1940s to the 1970s was a period of rapid global industrialization and growth in fossil fuel use, especially coal and oil. This period also preceded most regulation and control of particulate air pollution from industrial and transportation sources. The level of reflective aerosol pollution (known as global dimming) was so high, especially in the form of sulfur dioxide from coal plants and tailpipe emissions from millions of inefficient internal combustion engines, that the warming impact of rapidly increasing concentrations of greenhouse gases was temporarily offset.[‡] The Clean Air Act in the United States, similar policies that reduced air

[*] Alexandra Witze, "Earth: 2010 ties record for warmest year: 20th century's global average exceeded for 34th year in row," *Science News* 179, no. 4 (2011).
[†] Witze, "Earth: 2010 ties record."
[‡] Flannery, *The Weather Makers.*

pollution in other industrial nations, and continued increases in atmospheric greenhouse gas concentrations were all factors that led to the resumption and acceleration of warming since 1976. High rates of sulfur emissions related to Asian industrialization and rapid construction of coal-fired electricity plants may have again slowed the rate of warming from about 2000 through 2010.[*]

Figure 1.3 shows global land–ocean temperature trends since 1880 in the left portion of the figure, and the 2008 surface temperature anomaly in comparison to the base period of 1951–1980 on the right. The rapid global warming since about 1980 is evident in the land–ocean temperature trend line. The 2008 surface anomaly illustrates the now typical pattern of greater warming at high latitudes. Although 2008 was the coolest year since 2000, it was among the ten warmest years in the instrumental record. The relatively cool temperatures in much of the Pacific Ocean that year were due to a strong La Niña during the first half of the year.[†] I return to La Niña and other natural influences on climate variability below.

Tim Flannery has noted that global warming tends to cause climate change in "jerks" as climate patterns move rapidly from one relatively stable state to another. He cites evidence supporting the idea of climatologist Julia Cole that the Earth passed through a "climatic magic gate" in 1976 and again in 1998. These two years of exceptional weather and the rapid rise in temperatures since the 1970s appear to be related to warming-induced changes due to the El Niño–La Niña cycles that are evident in higher surface water temperatures in the tropical Pacific Ocean.[‡] Time will tell if 1976 truly marked a long-term climactic watershed and if the rapid global warming shown in Figure 1.3 since the 1970s continues.

When thinking about perceptions of warming in the context of global averages, it is helpful to remember that the continental United States, for example, constitutes about 1.5% of the Earth's surface area. An unusually cool summer or cold winter on one continent, or in a region, state, or local area does not serve as an informed basis to gauge global averages. Measured warming has been greater in the Northern Hemisphere than the Southern Hemisphere, greater over land than over the oceans, greater at the Poles, and greater at higher elevations than in lower-lying regions. Weather anomalies such as drought, heavy rains and snowfall, flooding, extreme cold snaps, and intense storms are occurring with greater frequency on every continent.[§] The insurance industry is acutely aware of the increasing frequency of damaging weather and climate-related disasters. The changes affecting the insurance industry are not limited to higher claims due to storms damaging high-value property in developed areas. Since 1980 there has been a steady

---

[*] Nadia Drake, "Environment: sulfur stalled surface temperature: coal emissions explain why warming stopped for a decade," *Science News* 180, no. 3 (2011).

[†] Goddard Institute for Space Studies, "Global Temperature Trends: 2008 Summation," http://www.data.giss.nasa.gov/gisstemp/2008

[‡] Flannery, *The Weather Makers*, 84–85.

[§] United Nations Environment Programme, *Climate Change Science Compendium 2009*.

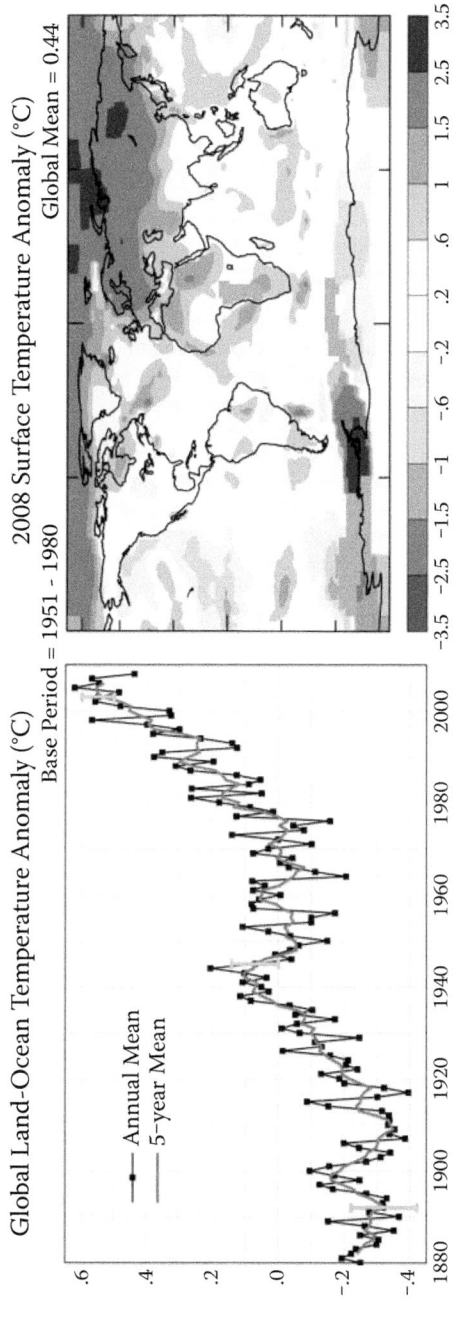

FIGURE 1.3 (See color insert.)
Global Land–Ocean Temperature Anomaly. *Source:* National Aeronautics and Space Administration, Goddard Institute for Space Studies.

global increase in the incidence of damaging storms with high winds, floods, droughts, and wildfires.[*]

1.3.1 Natural Factors and Denial

The analysis of climate change by GISS and other climate scientists includes consideration of both the natural and anthropogenic influences on climate. It is important to have a basic understanding of natural variation, cycles, and long-term climate changes in order to appreciate the discontinuous nature of anthropogenic climate change. It is also helpful to be aware of natural factors and their documented impacts in order to make informed judgments about the claims made by climate change deniers or "contrarians" as aptly named by GISS scientist James Hanson. I will not devote a lot of space to the arguments of contrarians, but some consideration is relevant and useful. Their central arguments continue to gain attention despite the fact that they have been thoroughly discredited in the scientific literature.[†] James Hanson has succinctly elaborated on the countervailing evidence for several contrarian claims.[‡] The evidence of measured warming and the central role of anthropogenic greenhouse gases are no longer in credible scientific dispute. Some of the more prominent deniers with scientific credentials have recently reviewed and confirmed the validity and authenticity of global warming evidence without addressing the more complex issue of anthropogenic causes.[§]

One recent and instructive example of evidence correcting contrarian claims was published in *Science* by Andrew Lacis and his colleagues in 2010. Lacis et al. directly addressed the popular deniers' claim that water vapor, rather than CO_2, is responsible for the great majority of the greenhouse effect that is warming the planet. Lacis et al. demonstrated that CO_2 and other persistent noncondensing greenhouse gases like CH_4 and N_2O provide a stable temperature structure that sustains current levels of water vapor and clouds in the atmosphere. Water vapor and associated cloud feedbacks account for about 75% of the greenhouse effect, but "without the radiative forcing of CO_2 and the other noncondensing greenhouse gases, the terrestrial greenhouse would collapse, plunging

---

[*] Janet Raloff, "Insurance payouts point to climate change," *Science News*, Web Edition (2012), http://www.sciencenews.org/view/generic/id/337318/title/Insurance_payouts_point _to_climate_change

[†] Naomi Oreskes and Erik M. Conway, *Merchants of Doubt: How a Handful of Scientists Obscured the Truth on Issues from Tobacco Smoke to Global Warming*, 1st U.S. ed. (New York: Bloomsbury Press, 2010); James Lawrence Powell, *The Inquisition of Climate Science* (New York: Columbia University Press, 2011).

[‡] James E. Hansen, *Storms of My Grandchildren: The Truth About the Coming Climate Catastrophe and our Last Chance to Save Humanity*, 1st U.S. ed. (New York: Bloomsbury USA, 2009).

[§] Seth Borenstein, "Skeptic finds he now agrees that global warming is real," *Lawrence Journal World*, October 31, 2011, 6A.

the global climate into an icebound Earth state."[*] The greenhouse effect maintains the average global surface temperature of the Earth at about 59°F (and rising). Without greenhouse gases the average surface temperature on the Earth would be about −4°F.[†] Lacis et al. provide evidence that the triggering mechanism, or "control knob," for the greenhouse effect and contemporary global warming is CO_2 and the other noncondensing greenhouse gases.[‡]

This example of evidence contradicting or clarifying a claim made by deniers is instructive because it elucidates some of the central features of such claims. Those features include attributing documented global warming to a natural factor or cycle and a failure to account for anthropogenic changes to the composition of the atmosphere and the carbon cycle. Water vapor is a natural and large part of the Earth's atmosphere. It plays a central role in the greenhouse effect. Both are factual statements. As the Earth and its troposphere retain more heat, the volume of water vapor in the troposphere also increases, which affects cloud formation and precipitation patterns. Why is there additional heat in the troposphere beyond what can be explained by natural factors? Evidence supporting the central role of anthropogenic greenhouse gases is robust and unequivocal.

1.3.2 Uncertainties Remain

There are many remaining scientific uncertainties regarding climate change. The influence of aerosols as factors in both cooling and warming remains a central focus of research. There also remains much to learn about the local impacts of global warming. Global temperatures are projected to increase an additional 2°F–11.5°F by 2100, but local effects could be much stronger or weaker.[§] Climate researchers have robust evidence that, on a global scale, temperatures will continue to rise due to a strengthening greenhouse effect. Warming is expected to continue to be more extreme at northern latitudes than in other areas of the globe. Precipitation is expected to increase at high latitudes and decrease at low latitudes. When asked to provide temperature or precipitation projections for a specific location or region, however, the models that can summarize global trends relatively well are not reliable.[¶] Extreme heat waves and droughts will likely continue to occur with greater frequency, but their timing and location are not known with precision. Future changes in precipitation and extreme precipitation events, the exact behavior of sea ice and continental ice shelves, and the impact of warming on ocean currents are all uncertain and important. The rate and degree of future sea

---

[*] Lacis et al., "Atmospheric CO2: principal control knob, 356.
[†] Flannery, *The Weather Makers*.
[‡] Lacis et al., "Atmospheric CO2: principal control knob.
[§] IPCC, *Climate Change 2007*.
[¶] Richard A. Kerr, "Vital details of global warming are eluding forecasters," *Science* 334, no. 6053 (2011).

level rise will be closely related to the rate of melting glaciers and ice shelves on Greenland and Antarctica, which remain imprecisely understood.

Two additional and related scientific uncertainties are particularly important to the central focus of this book. Carbon cycle feedbacks and changes in land use are both topics of great uncertainty and therefore are not well incorporated into models of future climate. Carbon cycle feedbacks are complex and changing. Uncertainties remain regarding exactly how carbon cycles through the atmosphere, ocean, and land systems. It is well known, for example, that rising temperatures melt permafrost and release carbon into the atmosphere. The amount of carbon released is uncertain. Higher temperatures and less snow and ice cover in the areas affected by permafrost melt are also causing increases in biomass. As more and larger plants grow on melting permafrost, some of the carbon that enters the atmosphere from the permafrost is offset by that which is sequestered in plant tissues. The relative permanence of that sequestration as well as the net magnitude of carbon released are unknown and the topic of ongoing research.[*] To complicate matters further, plants growing in areas previously snow covered absorb more solar energy and reflect less heat (lower albedo), and thus add to the net warming effect.

Land use change and its impact on the carbon cycle and albedo is another area of uncertainty that is not well incorporated in climate models. Clearing forest land for agriculture and degrading grasslands by overgrazing reduce the carbon stored in plants and in the soil. Most of this carbon is released into the atmosphere in the form of CO_2 and CH_4 and is generally believed to account, as previously noted, for about 20%–25% of anthropogenic greenhouse gas emissions.[†] The microbial communities in soil are altered by changes in temperature and by agricultural chemicals in ways that are also not fully understood. Some research has shown that warming soils store less and release more carbon as they warm. Most climate models assume that the land system will be less of a carbon sink and possibly a net carbon source for the atmosphere as global warming continues. Recent research has shown, however, that the release of carbon from the soil as the environment warms is not a simple linear process. Microbes, bacteria, and fungi in the soil adapt to warming temperatures. Their metabolic rate and the mix of species change as the temperature changes. The specific characteristics of different soils, moisture levels, and plant populations affect the impact of warming on carbon release. Perhaps the greatest uncertainty related to land use is that it can and does change. We collectively can slow the historical degradation of the land system. The magnitude and impact of such action is an open question and a source of great uncertainty.

---

[*] J. Richter-Menge, and J. E. Overland, eds., "Arctic Report Card 2010," http://www.arctic.noaa.gov/reportcard

[†] Dawson and Spannagle, *The Complete Guide to Climate Change.*

1.3.3 The Relative Impact of Natural and Anthropogenic Factors

Before looking at the evidence regarding the relative strength of the natural and anthropogenic factors on measured warming, a brief overview of the key sources of natural climate variation is useful. The El Niño Southern Oscillation (ENSO) cycle in the tropical Pacific Ocean is the dominant natural interannual climate variation on Earth. Measurable variations in average regional and global temperatures are caused by the tropical El Niño–La Niña cycles. The name El Niño (the Christ child in Spanish) refers to the warm ocean current that periodically occurs, usually around the end of December, off the Pacific coast of Peru and Ecuador. The La Niña period is characterized by an upwelling of cold water in the same area that spreads west through the equatorial Pacific Ocean and lowers surface water temperatures. The cooler air temperatures that accompany the La Niña period are replaced with warmer ocean surface and air temperatures during the El Niño phase of the cycle. This naturally occurring cycle creates a two–seven year fluctuation of ocean temperatures and circulation patterns in the Pacific.*

The ENSO is an important example of the complex linked interactions between oceanic cycles and atmospheric cycles. The alternating changes in ocean and air currents in the South Pacific linked to ENSO can account for approximately 0.2°C or more of temperature variation in global average temperature. Even the ENSO, the strongest of the "natural" climate forcings, has been altered and intensified by anthropogenic global warming. Since 1976 the cycles have been unusually long—so long that it would be expected that such long cycles would occur once in several thousand years—and there has been an imbalance between the phases, with more La Niña and fewer and stronger El Niño cycles.[†]

Solar irradiance is an additional natural influence on climate variability. Solar irradiance, the amount of solar radiation reaching the Earth's atmosphere, fluctuates slightly on approximately an eleven-year cycle. Warmer surface temperatures tend to occur during active portions of the solar cycle, known as solar maximums, and cooler temperatures accompany less active periods called solar minimums. Solar energy given off by the sun can increase or decrease by approximately 0.1% between solar maximums and solar minimums, which produces about 0.1°C (0.18°F) of cooling or warming.[‡]

Changes in solar radiation were once considered a potential explanation for a large portion of recent global warming. Evidence to support that theory has not emerged despite extensive research. The energy reaching the Earth from the sun has been measured precisely in recent decades by satellites. The sun's output has not increased since 1978, so the warming of the past thirty

---

* Dawson and Spannagle, *The Complete Guide to Climate Change.*
† Goddard Institute of Space Studies, "Research News," January 21, 2010, http://www.giss. nasa.gov/research/news
‡ Goddard Institute of Space Studies, "Research News," January 21, 2010.

years is not attributable to an increase in solar energy reaching the Earth.[*] Cyclical variations between solar maximums and solar minimums tend to cancel one another out. There is evidence, however, of a small but persistent increase in solar irradiance since preindustrial times that has contributed a small increase of about 0.12 watts per square meter of radiative forcing.[†] We will look at this natural warming factor in the context of other natural and anthropogenic sources of radiative forcing later. Radiative forcing refers to factors that change or "force" either incoming solar radiation or outgoing infrared radiation away from their natural state of balance.

Another natural factor that can have an impact on radiative forcing and global temperatures is volcanoes. Volcanoes are a large and powerful source of sulfate aerosols (particles and/or droplets) that reflect incoming solar radiation back into space. A major volcano like the 1991 eruption of Mount Pinatubo in the Philippines can cause a dip in global average temperature of about 0.3°C. Since the sulfate and other aerosols emitted by volcanoes mostly settle out of the atmosphere in a few years, their cooling effect is relatively short-lived.[‡] Over longer time periods volcanoes are a source of CO_2 emissions and have a small positive (warming) impact on climate.[§]

Aerosols introduced into the atmosphere on a continuous basis by human activities have a more persistent climate effect than those emitted by volcanoes. Up to 50% of the anthropogenic global warming that otherwise would have occurred due to higher atmospheric concentrations of greenhouse gases may have been thus far offset by a wide range of human-introduced aerosols.[¶] It may be one of the most brutal of all paradoxes that one type of air pollution may be reducing and delaying the severity of the impact of another form of deadly air pollution. The particulate pollution released by coal-fired power plants, cement manufacturing, petroleum refineries, agriculture, and the exhaust of all the internal combustion engines in the world is reflecting some solar radiation back into space and is thus offsetting, through global dimming, some of the warming caused by the greenhouse gases released by the very same sources.

This might be a happy outcome if these particulate emissions were not also the cause of a wide range of human health and ecological problems. Human activity is the source of one-quarter to one-half of all aerosols in the atmosphere.[**] Aerosols containing mercury, uranium, and other harmful elements from coal-powered electric plants accounted for about 60,000 deaths per year during the 1990s in the United States alone. Sulfur dioxide from these same coal plants killed trees and acidified lakes, especially at high latitudes in the Northern Hemisphere. Acid rain resulting from sulfur dioxide

[*] Weart, *The Discovery of Global Warming.*
[†] Dawson and Spannagle, *The Complete Guide to Climate Change.*
[‡] Goddard Institute of Space Studies, "Research News," January 21, 2010.
[§] IPCC, *Climate Change 2007.*
[¶] Hansen, *Storms of My Grandchildren.*
[**] Weart, *The Discovery of Global Warming.*

emissions from coal-fired power plants has since been greatly reduced by burning lower sulfur coal and through the adoption of cleaner technologies.*

There is also strong evidence that the direct reflective properties of aerosols that cause global dimming are not the full story of their impact on climate. Aerosols, both natural and man-made, also affect cloud formation and have a significant impact on precipitation. The net effect of changing cloud feedbacks in a warming climate is one of the greatest uncertainties in contemporary climate research. Most analyses of the effect of cloud feedback to date have shown that clouds have a net negative (cooling) effect because they reflect more solar energy than they trap. There is some evidence, based on ten years of observations from 2000 to 2010, that as the climate warms cloud changes will result in a net positive (warming) feedback.[†]

Not all aerosols of human origin have a direct or indirect cooling effect. Black "soot" or black carbon aerosols absorb sunlight and cause warming. Black soot is produced by diesel engines and stoves that burn coal, animal wastes, and field residue. It is particularly prevalent in China, India, and other developing nations and is one of the most dangerous aerosols to human health. Black soot has been recently recognized as a powerful source of radiative forcing that is exceeded only by CO_2 and CH_4. [‡]

1.3.4 Long-Term Cycles and Rate of Warming

The Earth's climate is maintained in a relatively stable state by the balance between incoming ultraviolet solar radiation and outgoing infrared radiation. The Earth–atmosphere system, when unperturbed by human action, is considered stable in timescales of a few centuries. Over longer, paleoclimatic timescales, changes occur due to factors such as continental drift and the Milankovich cycles. Significant continental drift occurs over eons and cannot, of course, explain the recent and rapid warming that has been documented by climate scientists.

Named after Serbian astronomer Milutin Milankovich, the Milankovich cycles refer to slight changes in the Earth's orbit and the tilt of the Earth's axis in relation to the sun. The Milankovich cycles occur over thousands of years and are believed to be a major factor in the initiation of glacial and interglacial periods. The three principle cycles in the Earth's orbit and orientation toward the sun described by Milankovich in 1941 are interesting and important in their own right and serve as a good window into the nature and complexities of climate change, both natural and man-made.

The longest of the Milankovich cycles involves changes in the Earth's orbit around the sun. The orbit varies from a near circular path to a more

[*] Flannery, *The Weather Makers*.
[†] A. E. Dessler, "A determination of the cloud feedback from climate variations over the past decade," *Science* 330, no. 6010 (2010).
[‡] Hansen, *Storms of My Grandchildren*.

elliptical or elongated pattern on a timescale of approximately 100,000 years. When the Earth is in its nearly circular orbit, as it is currently, the amount of solar energy hitting the Earth varies by approximately 6% between winter and summer. When the Earth's orbit is at its most elliptical extent, the solar energy reaching the planet varies by as much as 20%–30% from January to July. Since the Earth is carried both closer to and further away from the sun, the intensity of the solar energy reaching the planet varies substantially during the year. This is the only one of the three cycles that changes the total amount of the sun's energy that reaches the Earth, so it can have a substantial impact on climate.[*]

The second of the Milankovich cycles involves the tilt of the Earth's axis and occurs about every 42,000 years. The tilt of the axis affects where solar radiation hits the surface of the planet. In a complete cycle the tilt of the Earth's axis ranges from 21.8 to 24.4 degrees. We are now at about the middle of this cycle, with an axis tilt of approximately 23.5 degrees. At the greatest angle the difference between winter and summer temperatures is greatest. As the angle slowly decreases, as it is currently, the difference between winter and summer temperatures also decreases. Over the next 10,000 years the average temperature variation between summer and winter should be slightly reduced.[†]

The third, and shortest, of the Milankovich cycles takes about 22,000–23,000 years and involves the wobble of the Earth on its axis. During this cycle the Earth's axis shifts from pointing to the Pole star to pointing to the Vega star. This affects both the timing and intensity of the seasons. In about 10,000 years summer will be centered in January and February in the Northern Hemisphere instead of July and August.[‡] When the wobble in the Earth's axis is oriented so that Vega marks true north, summers can be extremely hot and winters very cold. If the latter occurs at the same time that the angle of the tilt of the Earth's axis is at or near its maximum of 24.4 degrees or near its minimum of 21.8 degrees, the two cycles can magnify or reduce their respective impacts on climate.[§]

At their most extreme configuration, Milankovich cycles can cause an annual variation of less than 0.1% in the total solar energy reaching the planet. This, however, is enough to trigger a series of changes that can lead to a variation in average global temperature of up to 5°C (9°F). Ice core records show that this synchronization may occur about every 120,000 years. These conditions are thought to create periods during which there is insufficient solar radiation in the northern latitudes to melt the previous winter's snow and ice cover. If the Northern Hemisphere includes the

---

[*] Dawson and Spannagle, *The Complete Guide to Climate Change*; Hansen, *Storms of My Grandchildren*.
[†] Dawson and Spannagle, *The Complete Guide to Climate Change*; Hansen, *Storms of My Grandchildren*.
[‡] Dawson and Spannagle, *The Complete Guide to Climate Change*.
[§] Flannery, *The Weather Makers*.

majority of the Earth's land mass, and that mass is near the North Pole, as it has been in the configuration of the continents that has existed during the past several glacial periods, the perennial accumulation of snow and ice in northern latitudes can substantially raise the Earth's albedo. A higher albedo increases the proportion of solar radiation reflected back into space, fundamentally altering the Earth's energy balance and magnifying the impact of the Milankovich cycles. This process is believed to have contributed to the onset of ice ages.[*] All of the factors initiating glacial and interglacial periods are still not completely understood, but "it is certain that greenhouse gases play a role" in the onset of the ice ages and the eventual thaw.[†] As previously noted, it is now robustly documented that greenhouse gases, especially CO_2, are the trigger or "control mechanism" of past and current climate change.

Uncertainties remain regarding the complete and precise sequence of events that led to each of the ice ages during the 1.8 million years of the Pleistocene epoch and the emergence of the 11,500 year summer of the Holocene. The quest to understand and explain the puzzle of the glacial cycles and the relative stability of the current long summer has been a central driver in the development of modern climate science and has contributed to an exponential growth of physical evidence and knowledge.

I noted earlier that a 0.1% variation in solar energy reaching the Earth due to a convergence of the Milankovich cycles has probably triggered a variation of up to 5°C (9°F) in average global temperature and is widely believed by paleoclimatologists to have led to the initiation of new glacial and interglacial periods. If a long-term natural cycle can produce such a large temperature variation, what is unique about the current warming? The short answer is the rate of change. Until now the fastest documented rise in overall surface temperature in recent Earth history was about 5°C (9°F) over a period of 10 millennia from about 20,000 years ago to 10,000 years ago.[‡]

Over the past three decades the average global surface temperature has increased by 0.2°C (0.36°F) per decade. The recent rate of increase translates to a rate of temperature change of 2°C (3.6°F) per century. The earlier change at the end of the last ice age, 5°C in 10,000 years, translates to a rate of 0.05°C (0.09°F) per century. The current rate of change of 2°C per century is about forty times faster than the previously recorded period of most rapid change. If a midrange projection of an additional warming of 4°C (7.2°F) by the end of this century proves to be accurate, that would constitute a rate of change eighty times faster than what is known to have occurred at the end of the last ice age.

---

[*] Dawson and Spannagle, *The Complete Guide to Climate Change.*
[†] Flannery, *The Weather Makers,* 42.
[‡] Dawson and Spannagle, *The Complete Guide to Climate Change*; Flannery, *The Weather Makers.*

In a research paper published in January 2011 and posted early online to coincide with the United Nations Framework Convention on Climate Change in Cancun, Mexico, in December 2010, Richard Betts, of the UK's Meteorological Office in Exeter, projected that we could experience a 4°C warming as soon as the 2060s. Betts based his projection on the greenhouse gases industrial nations are currently emitting and the near-term expected emissions of rapidly industrializing nations like India and China. Acknowledging that the key to future warming is the cumulative emissions of greenhouse gases, Betts noted that emissions will have to peak within five years (from 2011) if humans are to avoid increases exceeding 2°C above preindustrial levels.[*]

Regardless of which of the many scenarios for the future is more accurate, the current rate of increase in measured average global temperature is much faster than at any time recorded in recent Earth history. Recent Earth history as used here refers to the past two million years as documented in ocean sediments. The rate of warming is perhaps as important as the magnitude of total warming. Successful adaptation by plant and animal species is much more possible when changes occur over hundreds or thousands of years than when they occur over decades. In time frames relevant to human history and to the plant and animal species currently living on the Earth, the rate of the recent episode of warming is unprecedented.

1.4 CO$_2$ and Radiative Forcing

As summarized previously, there is an overwhelming international scientific consensus that the documented increase in average global temperatures during the industrial age is primarily due to the increased concentration of greenhouse gases in the atmosphere. The concentration of these gases remained relatively stable for 10,000 years and then increased by more than 40% since 1750.[†] The availability of reliable data on the atmospheric concentration of CO$_2$ and its rate of increase fundamentally altered scientific discourse on climate.

In the 1950s climatologist Charles Keeling began recording a series of measurements of the concentration of CO$_2$ in the atmosphere on top of Mt. Mauna Loa, Hawaii. Characterized as the *Silent Spring* of climate science by ecologist, evolutionary biologist, and author Tim Flannery, what has become known as Keeling's curve provided the key evidence needed to establish that humans were fundamentally altering the Earth's atmosphere. According to Flannery the consistent upward trend of Keeling's curve was the "first definitive sign that the great aerial ocean might prove to be the Achilles' heel of our fossil-

---

[*] Goddard Institute of Space Studies, "Research News," January 21, 2010.
[†] IPCC, *Climate Change* 2007.

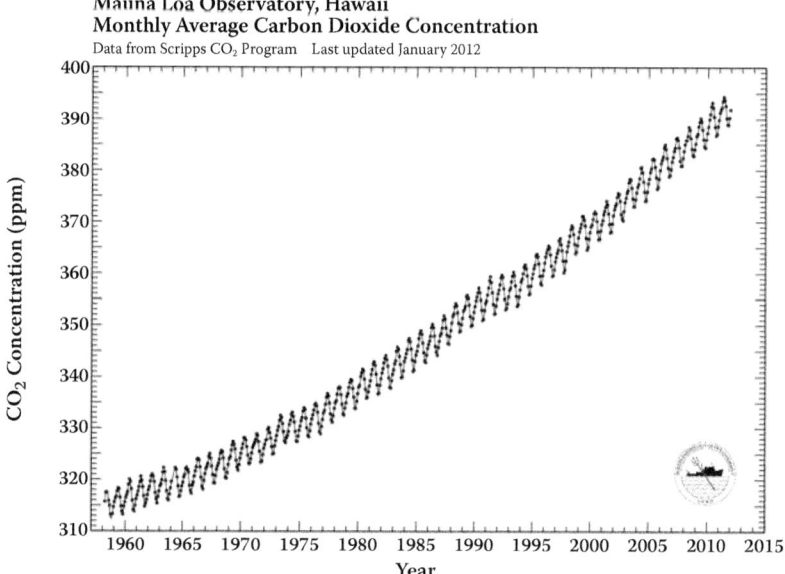

Mauna Loa Observatory, Hawaii
Monthly Average Carbon Dioxide Concentration
Data from Scripps CO₂ Program   Last updated January 2012

FIGURE 1.4
Keeling Curve. *Source*: Scripps CO₂ Project.

fuel-addicted civilization.'"* It now seems clear that the atmospheric capacity to absorb human wastes is indeed the weak link in a remarkably resilient ecosphere. More specifically, the physical process of warming set in motion by the chemical alteration of the atmosphere caused by human emissions of greenhouse gases is threatening to alter the Earth's climate to such a degree and at such a rapid pace that adaptation is very likely to become increasingly problematic for humans and many other species.

The oscillating sawtooth pattern of the Keeling curve shows the seasonal fluctuation in atmospheric CO_2 due to seasonal uptake and release of CO_2 in Northern Hemisphere forests and grasslands. Remember that the Northern Hemisphere includes the majority of the land surface on Earth and thus has more photosynthetic production than the Southern Hemisphere. A similar but smaller oscillation is evident in the Southern Hemisphere. To observe the seasonal oscillations captured in Keeling's measurements is to essentially witness the respiration of the planet. As plants grow and add leaves in the northern spring and summer the atmospheric levels of CO_2 decline by about 5–6 parts per million. This seasonal biospheric flux reverses in the fall and winter as plants lose their leaves and return much of the carbon taken up through photosynthesis back into the atmosphere. Each year the high point is a little higher. Keeling's measurements have been replicated across the

---

* Flannery, *The Weather Makers*, 25–26.

globe and the trend is the same everywhere.[*] More carbon is added to the atmosphere every year than can be taken up by the ocean and land systems. I will return to this issue in Chapter 2 and provide detailed information about how the carbon cycle functions, how climate change is affecting the natural cycle, and why I think careful treatment of the land-based carbon reservoir may be one key to successful mitigation of and adaptation to the imbalance so clearly illustrated by Keeling's curve.

The atmospheric concentration of CO_2 is increasing every year, and the rate of increase has accelerated in recent years. During the 10,000 years prior to 1750, ice core data show that concentrations of CO_2 fluctuated between approximately 270 parts per million and 290 parts per million, with a midpoint of 280 parts per million. In 2011 the atmospheric CO_2 concentration was 392 parts per million.[†] During the five years from 2000 to 2005 the atmospheric concentration increased by the same amount as it did during the first 150 years following the Industrial Revolution. Between 1958, when Keeling began his systematic measurements, and 1975 atmospheric concentrations were increasing at a rate of 9.4 parts per million per decade. Since 2000 the rate of increase has been 20 parts per million in one decade.[‡]

1.4.1 Human Impacts Recently Overwhelmed Natural Factors

There are several natural factors that have had and continue to have important impacts on climate change and variability. The ENSO, increased solar irradiance, volcanoes, and the Milankovich and other natural cycles were likely the dominant climate drivers in Earth's Holocene epoch. Using air bubbles trapped in ice, scientists have documented probable human impacts on the composition of the atmosphere dating back at least 8,000 years. These first detectable signs of global human influence in the form of increased CO_2 and CH_4 emissions from agriculture were of uncertain and debatable importance. Human land management practices, especially wet agriculture, altered the carbon cycle and may have made a small contribution to the unusually stable warm climate that has provided a very hospitable environment for the development of human civilization.[§]

For thousands of years, human agriculture, deforestation, and associated changes to the planet and its atmosphere were likely minor climate factors in comparison to natural factors and cycles. The concentration of atmospheric CO_2 began a modest assent in the early years of the Industrial Revolution, in about 1750, as coal came into increasingly widespread use. Even the first 200 years of the Industrial Age, however, and the development and spread of fossil fuel-based energy systems and industry did

---

[*] IPCC, *Climate Change* 2007.
[†] Vince, "An epoch debate."
[‡] Dawson and Spannagle, *The Complete Guide to Climate Change.*
[§] Flannery, *The Weather Makers.*

not create a dominant climate forcing that clearly overwhelmed natural factors. The meteorologists and geologists of the early and mid-twentieth century who argued that human actions had minimal impact on global climate were likely correct, until about 1950. Although the physics of the greenhouse effect had been discovered and verified during the nineteenth century, reliable measurements of greenhouse gas concentrations in the atmosphere and global temperature increases were not generally available until Keeling's measurements began in the late 1950s.[*] Science is a skeptical process, and until reliable, replicable evidence was available and thoroughly tested, the relative impact of natural and anthropogenic climate forcings could not be determined.

James Hanson, climate scientist at NASA's GISS, explicitly notes that natural factors, particularly the increasing level of volcanic activity in the twentieth century, were "at least as large as the human-made climate forcing up through the middle of the twentieth century."[†] It was not until the past several decades that the rate of emissions and cumulative volume of CO_2 and other anthropogenic greenhouse gases in the atmosphere overwhelmed natural forcings from volcanoes and increasing solar irradiance. Wes Jackson, founder of The Land Institute, in Salina, Kansas, a plant breeder and thinker to whom we will return later, has provided a compelling way to understand how recent the largest part of the industrial explosion has been. Jackson illustrates the recent growth in oil consumption by noting that, as of 2010, a 22-year-old had "lived through over half of all the oil ever burned." He points out also that the recent "120-year period in population growth, by no mere coincidence, has come accompanied by a comparable increase in fossil fuel consumption," and concludes that "we stand at a moment of history unlike any other."[‡]

The Anthropocene epoch that we have just entered is like no other geologic age or period of human history because for the very first time the behavior of one species has begun to overwhelm natural biological, chemical, and geological processes on a global scale.

1.4.2 Radiative Forcing

The global warming potential (GWP) index is used by climate scientists to compare the radiative forcing impact of each greenhouse gas to the same amount of CO_2, which is the gas with the greatest radiative forcing impact because of its persistence and relative abundance. There are many anthropogenic greenhouse gases, at least twenty-six are tracked by the IPCC, and they all have different residence times in the atmosphere and varying physical

---

[*] Weart, *The Discovery of Global Warming*.
[†] Hansen, *Storms of My Grandchildren*, 8.
[‡] Wes Jackson, *Consulting the Genius of the Place: An Ecological Approach to a New Agriculture* (Berkeley, CA: Counterpoint Press, 2010).

properties. GWPs are used to compare the radiative forcing caused by emissions of different greenhouse gases and are expressed as a CO_2 equivalent by multiplying the amount of the greenhouse gas emitted by its GWP value. Radiative forcing is "an imposed perturbation (disturbance) of the planet's energy balance."[*] It is measured in watts per square meter.

Since each greenhouse gas has a different residence time in the atmosphere, GWPs vary according to the time horizon selected. CH_4, for example, has an expected atmospheric lifetime of twelve years and a GWP of twenty-one (actually CH_4's GWP has been revised to twenty-three, but since the twenty-one value is part of the Kyoto Protocol, it will be used in official measures until at least late 2012).[†] In simple terms, this means that CH_4 is a powerful greenhouse gas, but its long-term radiative forcing impact is reduced by its relatively short residence time in the atmosphere. An obvious implication of this information is that reductions of CH_4 emissions could reduce radiative forcing relatively quickly. CO_2 has a GWP of 1 (the base unit), but its average residence time in the atmosphere is 100 or more years. The persistence and volume of CO_2 emissions make it the most important greenhouse gas in terms of total radiative forcing. The GWP values and atmospheric residence times of twenty-six greenhouse gases are available in the 1997 IPCC report "Guidelines for National Greenhouse Gas Inventories."[‡] Documenting the GWP, residence time, and emissions volume of greenhouse gases is critically important, in part, because the data are essential for the development and documentation of future emissions reductions or increases by nations, regions, states, and other entities.

The combined positive radiative forcing effects of the long-lived greenhouse gases—CO_2, CH_4, N_2O, and the halocarbons—are the greatest in magnitude and have the highest level of scientific understanding of the factors included in Figure 1.5. Note that the range of uncertainty for each of the climate forcings in Figure 1.5 is marked as a range of possible variation. The mild cooling (negative forcing) effect of stratospheric ozone slightly offsets the warming effect of tropospheric ozone. Surface albedo has been increased, made more reflective, by land use changes and has a slight cooling effect as well. This is due to deforestation, as grassland, cropland, and human infrastructure have a higher albedo than the trees they replaced. Black carbon on snow, which absorbs light and heat, is shown as counteracting about half of the albedo-related cooling due to land use. This is an area of some debate, as research by James Hansen and his colleagues at GISS has found a much higher positive radiative forcing of 0.6 watts per square meter due to black carbon aerosols.[§] Both direct and indirect aerosol effects are shown as having a negative radiative forcing. Note also that the IPCC considers aerosol

---

[*] Hansen, *Storms of My Grandchildren*, 5.
[†] Dawson and Spannagle, *The Complete Guide to Climate Change*.
[‡] IPCC, *IPCC Guidelines for National Greenhouse Gas Inventories* (Geneva: IPCC, 1997), http://www.ipcc.ch/pub/reports.htm
[§] Hansen, *Storms of My Grandchildren*.

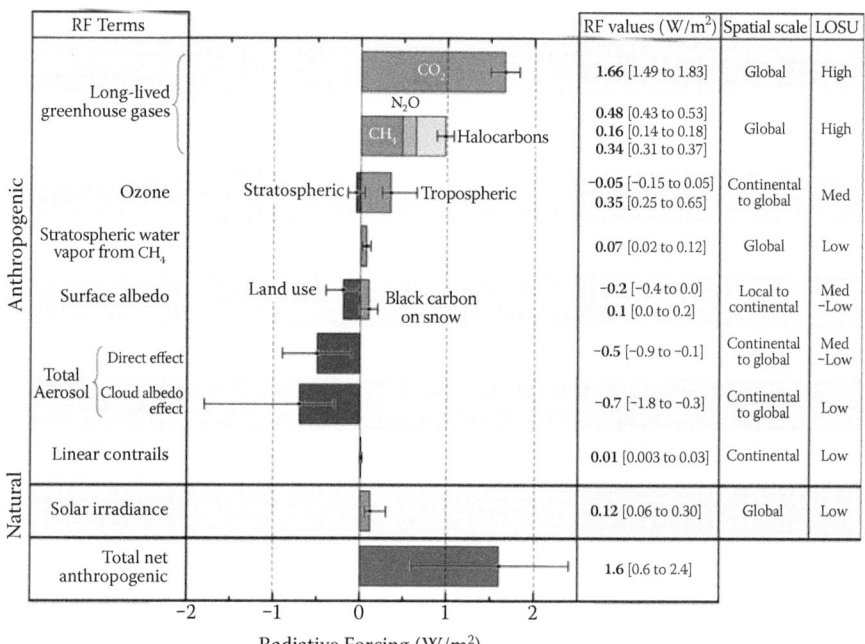

RF Terms	RF values (W/m²)	Spatial scale	LOSU
Long-lived greenhouse gases (CO₂, N₂O, CH₄, Halocarbons)	1.66 [1.49 to 1.83] / 0.48 [0.43 to 0.53] 0.16 [0.14 to 0.18] 0.34 [0.31 to 0.37]	Global / Global	High / High
Ozone (Stratospheric, Tropospheric)	-0.05 [-0.15 to 0.05] 0.35 [0.25 to 0.65]	Continental to global	Med
Stratospheric water vapor from CH₄	0.07 [0.02 to 0.12]	Global	Low
Surface albedo (Land use, Black carbon on snow)	-0.2 [-0.4 to 0.0] 0.1 [0.0 to 0.2]	Local to continental	Med -Low
Total Aerosol: Direct effect	-0.5 [-0.9 to -0.1]	Continental to global	Med -Low
Cloud albedo effect	-0.7 [-1.8 to -0.3]	Continental to global	Low
Linear contrails	0.01 [0.003 to 0.03]	Continental	Low
Solar irradiance	0.12 [0.06 to 0.30]	Global	Low
Total net anthropogenic	1.6 [0.6 to 2.4]		

Radiative Forcing (W/m²)

FIGURE 1.5 (See color insert.)
Radiative Forcing. *Source:* Intergovernmental Panel on Climate Change, 2007.

effects to have a low level of scientific understanding. The long-term increase in solar irradiance at 0.12 watts per square meter of positive radiative forcing is small compared to the 1.6 watts per square meter of net anthropogenic forcing. The 0.12 watts per square meter for solar irradiance shown in the IPCC data is the same as that reported by Hanson, but he includes a small positive forcing of about half that magnitude for CO₂ from volcanoes.[*]

As previously noted, Hansen argues that the evidence shows natural climate forcings from solar irradiance and volcanoes were not overwhelmed by human greenhouse gas emissions until the middle of the twentieth century.[†] Hansen's research demonstrates that the impact of the CO₂ effect on the Earth's radiation balance can be calculated accurately, with an uncertainty of less than 15%. Using data slightly more current than that presented above, Hansen concluded that the gross climate forcing due to CO₂ and the other anthropogenic greenhouse gases (CH₄, N₂O, chlorofluorocarbons, and ozone) was about 3 watts per square meter, with a net anthropogenic forcing very close to the value reported by the IPCC. The nature of the greenhouse effect and the resulting climate imbalance is succinctly summarized by Hansen in his 2009 book, *Storms of my Grandchildren*:

---

[*] Hansen, *Storms of My Grandchildren*, 6.
[†] Hansen, *Storms of My Grandchildren*, 6.

The largest human-made climate forcing is due to greenhouse gases. These are gases that partially absorb infrared (heat) radiation, so an increased gas amount makes the atmosphere more opaque at infrared wavelengths. This increased opacity causes heat radiated to space to arise from a higher level in the atmosphere, where it is colder. Heat radiation to space is therefore reduced, resulting in a planetary energy imbalance. So Earth radiates less energy than it absorbs, causing the planet to warm up.[*]

The 2007 IPCC Fourth Assessment Report found that there has been a total net positive change in anthropogenic radiative forcing of 1.6 watts per square meter since 1750. The IPCC concluded that this is most likely the largest sustained climate forcing in at least the past 16,000 years, and potentially the largest in the past several million years.[†] Those who argue that substantially warmer temperatures and higher concentrations of CO_2 have existed in the Earth's past can find supporting evidence only by looking at geologic periods prior to the evolution of humans and the ecological conditions upon which humans and other contemporary species depend.

1.4.3 The Paradox of Aerosols

It is apparent that the greatest uncertainties in the data presented above were associated with the impact of aerosols. This uncertainty continues in the scientific literature today.[‡] Hansen has presented evidence that both reflective aerosols and aerosol-related cloud changes are relatively large negative (cooling) forcings. In 2009 he cited a need for better data and acknowledged a high degree of uncertainty related to aerosols.[§] Since that time, strong evidence has emerged that the direct cooling effects of reflective sulfur particles were a factor in a lull in global temperature increases between 1998 and 2008.[¶] China doubled its coal consumption between 2002 and 2007, accounting for 77% of the rise in global coal use. Worldwide sulfur emissions rose by 26% during this period and effectively canceled out much of the warming effect of the rise in greenhouse gas emissions. Although this was part of the hottest decade in the instrumental record, the rate of temperature increases had stalled in relation to the decades-long trend. The relatively steady temperatures during this period were sometimes cited as evidence against global warming. The evidence, however, points to sulfur aerosol pollution temporarily counteracting the greenhouse effect.[**] This recent episode of global dimming from sulfur emissions—this time resulting from China's rapid

---

[*] Hansen, *Storms of My Grandchildren*, 5.
[†] IPCC, *Climate Change 2007.*
[‡] Witze, "The final climate frontiers."
[§] Hansen, *Storms of My Grandchildren.*
[¶] Drake, "Environment: sulfur stalled surface temperature."
[**] Drake, "Environment: sulfur stalled surface temperature."

industrialization—is reminiscent of the similar, but longer, warming hiatus from the 1940s to the mid-1970s, which coincided with the rapid industrial growth of the United States, Japan, and Europe.

New U.S. Environmental Protection Agency (EPA) rules begun in 2011 are designed to reduce both particulate and greenhouse gas pollution. China is making large investments to clean up particulate emissions from its coal plants, similar to the efforts in the United States following passage of the Clean Air Act in 1970.[*] The extent to which the industrial and developing nations are successful in reducing particulate air pollution will affect the degree to which greenhouse gas warming will continue to be partially offset by reflective aerosols. Because of the strong negative health and ecological effects, particularly respiratory disease and acid rain, of sulfur pollution, the effort to reduce them will likely continue. Unfortunately, success in reducing the damaging effects of sulfur dioxide pollution means accelerating the warming effect of greenhouse gases. Reducing the amount of coal burning would address both problems and reduce ocean acidification. The perplexing reality is that global coal consumption is increasing.

Yet another complicating effect of aerosols, and an example of the complex interactive nature of human ecological impacts, is the effect they have on CO_2 uptake by plants. Natural and man-made aerosols, including particulate matter and droplets of sulfuric acid and other pollutants from coal plants, transportation, and industry, scatter incoming solar radiation so that the light that reaches plants is more diffuse. Aerosols reduce the total amount of light that reaches a tree, for example, but the diffuse radiation actually illuminates more leaves below the tree's canopy. This scattered light enhances photosynthesis and increases the amount of carbon that the vegetation takes in. Short-term aerosol emissions from volcanoes have had the same effect of boosting plant productivity by diffusing light. Paradoxically, it has been projected that if particulate air pollution control measures improve atmospheric clarity, the increased natural biosequestration of carbon provided by diffuse light will nearly disappear by 2100.[†] Cleaner air has many health and ecological benefits, but will likely result in less carbon uptake by plants.

1.5 Climate Sensitivity: The Likely Extent and Rate of Warming

Climate sensitivity is the term used to describe the relationship between the various sources of radiative forcing (atmospheric greenhouse gas

---

[*] Drake, "Environment: sulfur stalled surface temperature."
[†] Sid Perkins, "Earth: aerosols may have boosted carbon uptake: plant productivity could drop as skies continue to clear," *Science News* 175, no. 11 (2009).

concentrations, changes in albedo due to melting polar ice, black soot, other human forcings, and natural factors) and changes in average global surface temperature. The IPCC uses a specific technical definition of climate sensitivity: the amount of global temperature change that would result from a doubling of atmospheric CO_2 concentrations above preindustrial levels (from 280 parts per million to 560 parts per million). Their 2007 Fourth Assessment Report concluded that a doubling of CO_2 above preindustrial levels would result in warming of 2°C–4.5°C (3.6°F–8.1°F), with the "most likely" increase of 3°C (5.4°F).[*]

The IPCC's midrange projections of increasing annual greenhouse gas emissions would most likely result in atmospheric greenhouse gas concentrations reaching 550 parts per million CO_2-eq by 2035–2040 and 630 parts per million CO_2-eq by 2050. These concentration levels would commit the Earth to an average temperature increase of 3°C–4°C (5.4°F–7.2°F), possibly higher, above preindustrial levels. Changes of this magnitude would "almost certainly result in dangerous climate change."[†]

Citing the wide range in IPCC temperature estimates and modeling uncertainties related to aerosol forcing and ocean heat uptake, Schmittner et al. recently combined temperature reconstructions from the Last Glacial Maximum (19,000–23,000 years ago) with climate modeling to estimate climate sensitivity. It was concluded that average temperatures will rise 1.7°C–2.6°C (3.1°F–4.7°F), with a mean estimate of 2.3°C (4.1°F), with a doubling of preindustrial CO_2 concentrations.[‡] This is the most conservative estimate of climate sensitivity to greenhouse gases found in the scientific literature and it projects dangerous climate change.

Dangerous climate change is often defined as a temperature increase of 2°C or more above preindustrial levels. Since we have already experienced an increase of 0.8°C, another 1.2°C (2.16°F) will cause us to cross this threshold. Note that the midrange projections of the IPCC, considered extremely conservative by many scientists,[§] have us likely crossing and exceeding this level in about 2035. A warming of this magnitude will cause significant climate changes, require extensive adaptation and result in serious and widespread ecological damage.[¶] Limiting greenhouse gases to a doubling of preindustrial levels and limiting warming to 2°C both appear increasingly unlikely in light of emissions trends since 2000. Stabilization of atmospheric CO_2 equivalents below 650 parts per million (560 parts per million is twice the preindustrial level) is now unlikely, and temperature

---

[*] IPCC, *Climate Change 2007*.

[†] Dawson and Spannagle, *The Complete Guide to Climate Change*, 186.

[‡] Andreas Schmittner et al., "Climate sensitivity estimated from temperature reconstructions of the last glacial maximum," *Science* 334, no. 6061 (2011).

[§] Dawson and Spannagle, *The Complete Guide to Climate Change*; Flannery, *The Weather Makers*; Hansen, *Storms of My Grandchildren*.

[¶] H. Schellnhube et al., eds. *Avoiding Dangerous Climate Change* (Cambridge: Cambridge University Press, 2006).

increases of 4°C are possible at that level of atmospheric greenhouse gas concentration.[*]

The term "dangerous climate change" is also a substitution for the term "dangerous anthropogenic interference" with the climate system as stated in Article 2 of the 1992 United Nations Framework Convention on Climate Change (UNFCCC). The definition of dangerous anthropogenic interference in the climate system was not quantified as a specific increase in global temperature by the UNFCCC. The convention did provide guidance that is very useful in thinking about the factors that would make climate change dangerous. The UNFCCC had as its primary objective to stabilize atmospheric greenhouse gas concentrations at a level and in a time frame sufficient to "allow ecosystems to adapt naturally to climate change, ensure that food production is not threatened, and enable economic development to proceed in a sustainable manner."[†] With the current rapid and accelerating rate of plant and animal extinctions, increasingly common local and regional climate-related food production shortages, and the still increasing volume of annual global greenhouse gas emissions, one could and many do argue that we have passed the threshold of dangerous climate change and are approaching catastrophe.[‡]

There has been a remarkable consistency among the projections, in both magnitude and range of uncertainty, that climate scientists and research organizations have made regarding climate sensitivity since about 1976. The highly similar midpoints of the projections may be due to the high degree of accuracy in understanding the climate forcing due to greenhouse gases, as noted by James Hansen. However, the wide range of possible temperature extremes is due to the many uncertainties related to aerosol reflectivity and changes over time, aerosol cloud impacts, and land and ocean carbon sink changes and feedbacks. The remarkable complexity of accounting for atmospheric and ocean circulation patterns and changes in those patterns as more heat, and thus more moisture, are added to the atmosphere also contribute to the wide range in possible temperature increases. There is, however, a clear consensus that the likely temperature anomalies are more likely to be at the high end of the projected ranges than at the lower end.[§]

The likely rate and extent of temperature change are, obviously, both critically important to our future ability to adapt. The IPCC and other research groups have developed several emissions scenarios that are based on varying

---

[*] Kevin Anderson and Alice Bows, "Reframing the climate change challenge in light of post-2000 emission trends," *Philosophical Transactions of the Royal Society A: Mathematical, Physical and Engineering Sciences* 366, no. 1882 (2008).

[†] Dawson and Spannagle, *The Complete Guide to Climate Change*, 112.

[‡] Flannery, *Now or Never*; Hansen, *Storms of My Grandchildren*; Elizabeth Kolbert, *Field Notes from a Catastrophe: Man, Nature, and Climate Change*, 1st U.S. ed. (New York: Bloomsbury, 2006).

[§] Flannery, *Now or Never*; Hansen, *Storms of My Grandchildren*; Elizabeth Kolbert, *Field Notes from a Catastrophe: Man, Nature, and Climate Change*, 1st U.S. ed. (New York: Bloomsbury, 2006).

assumptions regarding population growth, rates of technology development and deployment, and rates of economic growth. A detailed discussion of the various scenarios is beyond the scope of the present discussion. The interested reader is referred to the IPCC Special Report on Emission Scenarios (SPES), available online. There is also a rapidly developing body of literature analyzing cumulative emissions, the magnitude and rate of temperature change, and the likely impacts of specific temperature changes.[*]

The four groups or families of scenarios developed by the IPCC SPES do not incorporate assumptions about the development of any major emission mitigation policies that would alter greenhouse gas emissions. They do include varying degrees of renewable energy development and energy efficiency in addition to a wide range of population and economic growth rates and patterns. The SPES scenarios, originally published in 2000, were revised for the 2007 Assessment Report, but the overall projections remain similar. The large range of the IPCC possible outcomes, as distinguished from the most likely midrange projections, in my view, makes them primarily of value to demonstrate clearly that the future magnitude and rate of global warming depends on our collective actions from this point forward. I can think of no more important lesson.

1.5.1 The Uses and Limitations of Climate Models

The majority of the information provided to this point about global warming and climate change is based on direct measurements and observations by scientists and scientific organizations worldwide. We know from the instrumental record, for example, that the Earth's mean surface temperature, the mean air temperature in the lower troposphere, and the mean ocean surface temperature have all been rising since 1880, and there has been a rapid increase in the rate of that warming since about 1976, except for a ten year hiatus due to the mitigating effects of aerosol pollution. Similarly, knowledge of ocean acidification and the declining extent of sea ice, glaciers, and continental ice shelves is based on direct observation and measurements. Sea level rise and its increasing rate are also well documented.[†] A number of other documented climate-related changes such as biodiversity loss, earlier spring greening, migration of many plant and animal species to higher altitudes or toward the poles and the increasing frequency of droughts, floods, and other extreme weather events do not require projections into the future or the use of models. They are occurring now and are becoming increasingly well documented.[‡] In order to analyze and prepare for the likely magnitude

---

[*] Mark Lynas, *Six Degrees: Our Future on a Hotter Planet* (Washington, DC: National Geographic, 2008).

[†] J. Blunden et al., "State of the climate in 2010," S1–S266.

[‡] United Nations Environment Programme, *Climate Change Science Compendium 2009*.

and rate of future warming, however, it will be useful to utilize scientific projections derived from increasingly sophisticated climate models.

Early climate models were limited to primarily analyzing circulation patterns in the atmosphere. As climate scientists learned that the oceans and their currents were also central to understanding global climate, the models became increasingly complex and included both atmospheric and ocean currents. There are now about ten major global circulation models (GCMs) designed to simulate the behavior of the atmosphere and to predict changes in the future. Some of the most sophisticated models are developed, operated, and continually modified by teams of scientists at the Hadley Centre, Exeter, England; the Lawrence Livermore National Laboratory, Livermore, California; the Max Planck Institute for Meteorology, Hamburg, Germany; and NASA's GISS in New York City.[*]

There are two general types of equations included in a modern climate model. The first expresses fundamental physical principles, such as the laws of gravity and the conservation of mass and energy. The second type of equation sets parameters for patterns and relationships that have been observed in nature but may not be fully understood. There are several major criteria for evaluating the credibility of GCMs. Is the model consistent with the basic laws of physics? Can it accurately simulate the current climate? Can it simulate the ongoing changes in the weather systems that make up the current climate? Does the model accurately simulate what is known of past climates? All of the better computer models like those listed above do a reasonably accurate job of meeting these tests.[†]

The accuracy of climate models also depends on the accuracy and completeness of the empirical evidence included in the model. When Canadian researchers discovered in 2003, for example, that greenhouse gases affect sea level pressure as well as temperature, this new evidence had not previously been incorporated in GCMs. Prior to incorporating this evidence, climate models had underestimated the impact of climate change on North Atlantic storms.[‡] The same principle applies to any new discovery throughout the interdisciplinary and international community of climate science. As new evidence is discovered or modified it must be incorporated into the models. Judgments must continually be made by the scientists refining the models regarding the veracity of new findings and when new research findings are significant and reliable enough to justify model revision. The major uncertainties in contemporary climate science related to reflective aerosols, aerosol feedback impacts on clouds, and the future behavior of land and ocean carbon sinks all make the process of refining and correcting global circulation models mind-numbingly complex.

---

[*] Kolbert, *Field Notes from a Catastrophe*.
[†] Flannery, *The Weather Makers*; Kolbert, *Field Notes from a Catastrophe*.
[‡] N. P. Gillett et al., "Detection of human influence on sea-level pressure," *Nature* 422, no. 6929 (2003): 292–294.

Models are important to the ongoing development of climate science and are the best tools we have to get a reasonable estimate of the magnitude and rate of future warming and climate change. Ideally, credible estimates can be used by policymakers to gauge the urgency and scale of mitigation and adaptation strategies. Computer models are also used to obtain many of the measurements that scientists use. Converting satellite measurements of atmospheric radiances, for example, into observations of temperature involves the use of computer models.[*]

Climate change deniers often assert that climate models are unreliable and based on speculation rather than evidence. This criticism demonstrates a misunderstanding of how general circulation models work. The models now in use incorporate as much empirical data as possible to develop testable hypotheses about future climate change. Contrary to deniers' claims that models lead to unsubstantiated predictions of catastrophic climate change, recent observational evidence about past climate change indicates that models may be underestimating climate sensitivity, and particularly the positive feedback processes that can amplify global warming beyond current modeling estimates.[†]

Climate models have significant limitations because of the rapidly growing body of evidence that must be accounted for and the uncertainties associated with some important variables. There is ample evidence from direct observation, instrumental measurement, and computer models about the direction and approximate magnitude of future climate change to serve as the basis for action. It is important for the scientific teams working on climate models to continue to refine their databases and projections. As previously noted, current climate models are more useful and accurate in estimating general climate changes on a global scale than they are in making specific projections for specific locations. This is currently a serious limitation on the usefulness of modeling as a tool for guiding local climate change adaptation planning.[‡]

Despite the limited utility of climate models, mitigation and adaptation plans are in development and being implemented throughout the world. Many governments, nongovernmental organizations, corporations, and individuals are not waiting for more evidence before acting.[§] A key question is whether the response is commensurate with the rate and magnitude of emissions, ecological degradation, and climate change. With still accelerating global greenhouse gas emissions, widespread land management practices

[*] Paul N. Edwards, *A Vast Machine: Computer Models, Climate Data, and the Politics of Global Warming* (Cambridge, MA: MIT Press, 2010).

[†] Flannery, *The Weather Makers*.

[‡] Kerr, "Vital details of global warming are eluding forecasters."

[§] National Research Council, *Informing an Effective Response to Climate Change: America's Climate Choices*, ed. Board on Atmospheric Sciences and Climate (Washington, DC: National Academies Press, 2010).

that degrade soils, and intensifying climate change impacts, it seems clear that it is not.

One of the founders of ecological economics, Herman Daly, offered a succinct and penetrating argument for moving from the potentially paralyzing analysis and complexity of climate models to the relative certainty and willingness to act that comes from focusing on "first principles." Citing physicist John Wheeler, Daly observed that "we make the world by the questions we ask." Climate modelers ask unending questions about the projected atmospheric concentrations of CO_2, the date when a projected concentration will be reached, the specific increase in temperature that will result, the likely physical consequences, their sequence, and the costs of abating the damages. These questions are necessary for the advancement of science, but at the same time they create a world of "such enormous uncertainty and complexity as to paralyze policy." Writing in 2007, Daly argued that it was time to ask a different set of questions that may create a different world. Why not ask, "can we systematically continue to emit increasing amounts of CO_2 and other greenhouse gases into the atmosphere without eventually provoking unacceptable climate change?" Daly contends that scientists overwhelmingly agree that the answer is no. "The basic science, the first principles and the directions of causality are very clear." This question and its answer leads to a world of relative certainty that makes the basic direction of needed policy and action clear as well. Action must be taken on the basis of first principles. Focusing exclusively on second-order uncertainties leads to inaction.[*]

1.6 The Ecosystem Impacts of Global Warming and Related Ecological Crises

The geophysical and biophysical impacts of human actions on the Earth's systems during the Anthropocene epoch are increasingly dangerous, ubiquitous, and well documented. The loss of biodiversity, ocean acidification, declining freshwater availability, more frequent extreme weather events, stratospheric ozone depletion, expanding threats to human health, altered nitrogen and phosphorous cycles, rising sea levels and coastal zone impacts, aerosol pollution, chemical pollution, and the disruption of agriculture and food supplies comprise a partial list of the impacts of human actions that threaten the viability of human societies and the survival of millions of species on Earth.

---

[*] Herman E. Daly, "Climate Change: From 'know how' to 'do now'," Common Dreams.org May 13, 2008, http://www.commondreams.org/archive/2008/05/13/8925

1.6.1 Marine Ecosystems

Several of these impacts—declining freshwater availability, extreme weather events, altered nitrogen and phosphorous cycles, chemical pollution, and agricultural disruptions—have particularly direct implications for land management practices and food production. Ocean acidification and sea level rise are two effects of human actions that will be long lasting—at least several hundred and probably thousands of years—and are likely to have a strong negative affect on food supplies. Approximately one-quarter of humans' dietary protein is derived from marine environments.[*]

Overfishing and general overexploitation of marine organisms have severely reduced the abundance and health of marine ecosystems. These ecosystems support a vast range of phytoplankton, algae, and other photosynthetic organisms that form the base of the marine food chain. Zooplankton, fish, crustaceans, corals, marine mammals, amphibians, reptiles, and a range of other species are all threatened by ocean acidification. Marine ecosystems also support many terrestrial organisms, including sea birds and mammals. Already degraded marine ecosystems are now experiencing a fundamental alteration of their chemical composition comparable to the chemical alteration of the atmosphere. Like the energy/chemical imbalance in the atmosphere, the chemical imbalance in the oceans is from the same source—greenhouse gases, particularly CO_2. Pollution with plastics, agricultural and industrial chemicals, and hypoxia (loss of oxygen) from agricultural nutrient runoff exacerbate the universal acidification.

Like the surface temperature anomalies of global warming, the acidification and warming of the oceans is occurring at a higher rate and magnitude in polar areas than elsewhere. Cold water absorbs more CO_2 than warmer water and polar oceans are now more acidic than oceans at lower latitudes. Significant amounts of carbon are absorbed by the oceans and their absorptive capacity is diminished as they become warmer and more acidic.[†] Oceans vary in their capacity to absorb CO_2. In addition to temperature, absorptive capacity varies by the available supply of dissolved carbonate compounds needed to neutralize the carbonic acid formed when CO_2 is absorbed in surface waters. Carbonates flow into the oceans from the weathering of terrestrial limestone and chalk and are also derived from the chemical decomposition of the shells and skeletons of marine organisms. The North Atlantic basin and the shallow basin of the North Sea, between Great Britain and northern Europe, are together so effective at absorbing CO_2 that they have been called the "carbon kidney" of the Earth. The shallow and stratified waters of the North Sea are unusually efficient at absorbing CO_2, where currents transport it to the cold waters of the North Atlantic basin. This carbon kidney has only recently been discovered by oceanographers and it is now

---

[*] Dawson and Spannagle, *The Complete Guide to Climate Change.*
[†] Flannery, *The Weather Makers.*

feared that its effectiveness may be diminished by warming temperatures and ocean circulation changes initiated by climate change.[*]

Over the past century of higher atmospheric concentrations of CO_2, additional CO_2 uptake has lowered the average ocean pH (potential of hydrogen) level by 0.1 pH units, from 8.2 to 8.1. This represents a 25% increase in hydrogen ion concentrations. The IPCC projects that ocean pH could fall by an additional 0.14–0.35 points by 2100, depending on future emission rates. This would mean that ocean pH could reach as low as 7.75 and would represent a level of acidity not experienced in at least twenty million years.[†] Equally important and dangerous to a wide range of marine species is the "unprecedented rapidity of CO_2 release currently taking place," leading to a rate of ocean acidification believed to be more rapid than any episode over the past 300 million years.[‡]

The primary direct threat of increased ocean acidity is to calcifying marine organisms that use carbonate ions to form their shells and skeletons. These organisms (e.g., corals and planktons) are critical components of marine food chains and their decline will have profound and far-reaching ecosystem effects. A general decline in the abundance of calcifying organisms could also reduce the amount of carbon, in the form of shells and skeletons, sequestered on deep ocean floors. This means that the ocean carbon sink will likely be reduced in two ways, by lower rates of CO_2 absorption at the surface and less long-term sequestration in the deep oceans.[§] Oysters and other organisms that utilize aragonite for their shells are even more vulnerable to acidification than mollusks like crabs and prawns that use calcite. Aragonite is a calcium carbonate of different structure than calcite, which has a lower dissolution threshold in acidic conditions. The sub-Arctic North Pacific has a lower saturation point for carbonates than tropical oceans and thus is likely to be where oysters and similar organisms show earlier signs of carbonate deficit.[¶] These are just a few examples of the complex interactive ecological impacts resulting from human alteration of ocean chemistry.

Knowledge about the overall health of marine ecosystems is currently limited by what ecologists refer to as a "data deficit." During the past ten years 2,700 scientists from eighty nations have been compiling evidence for the first ever census of marine life. Since the year 2000 some 16 million records have been assembled into accessible databases, with 6,000 new species discovered. There are now about 240,000 known marine species, and it is estimated that at least one million exist. Much less is known about the other two key factors related to marine biodiversity—distribution and abundance. According to Ian Pointer, a marine ecologist who chaired the scientific steering committee

---

[*] Flannery, *The Weather Makers*, 33–34.
[†] IPCC, *Climate Change* 2007.
[‡] Bärbel Hönisch et al., "The geological record of ocean acidification," *Science* 335, no. 6072 (2012): 1058.
[§] Dawson and Spannagle, *The Complete Guide to Climate Change*.
[¶] Flannery, *The Weather Makers*.

for the marine census project, key findings were that marine life is more complex and interrelated than expected and that "humans have had far more impact on the oceans than we had imagined."* One more example of how human impact defines the Anthropocene.

Sea level rise will also have long-term ecological impacts. Upright growing corals will have difficulty keeping up with sea level increases in an effort to maintain optimum depth ranges. Add this to the bleaching from warmer water temperatures and the lower availability of carbonate ions for skeletal material due to acidification, and it is understandable why corals are in deep trouble. Coastal wetlands and wildlife breeding grounds will be lost to rising waters, and the degree to which they will be replaced by comparable new habitats is uncertain. Changes in storm surges, wave action, erosion and sediment deposition, and saltwater intrusion into freshwater aquifers and habitats, among other impacts, are all occurring in coastal areas around the globe.[†]

The IPCC estimated in 2007 that global mean sea level will rise 28–58 cm (11–23 inches) by 2100. Because of uncertainties regarding the rate of melt in the Greenland and Antarctic continental ice shelves, the largest ice masses on the planet were excluded from the IPCC estimates. Recent research, however, has documented an increasing rate of sea level rise and many believe that the IPCC estimates are likely conservative.[‡] The IPCC noted that from 1961 to 1990 sea levels rose an average of 1.8 mm per year and since 1993 the rate has increased to 3.1 mm per year. About half of sea level rise is attributable to thermal expansion and about half from melting of glaciers and other land-based ice. It is clear that because of the thermal inertia of the massive oceans it will take at least decades for the increase in global air temperatures that has and will occur to be fully realized in increased ocean temperatures and thermal expansion.

1.6.2 Terrestrial Ecosystems and Complex Ecological Interactions

Anthropogenic climate change and biodiversity loss both originate from many of the same human actions. The interrelated nature of the causes and consequences of ecological problems is a key feature of the dilemma we face and a key feature of the most effective strategies for mitigating those problems. The strategies to mitigate climate change also generally have the potential to mitigate biodiversity loss as well. Climate change is likely to displace habitat destruction as the most powerful future driver of biodiversity loss.[§] Strategies to improve the resilience of forests, grasslands, and soils have a positive impact on biological diversity and contribute to climate change

---

* Dennis Normile, "Counting the ocean's creatures, great and small," *Science* 330, no. 6000 (2010) p. 25.
† United Nations Environment Programmme *Climate Change Science Compendium 2009.*
‡ Dawson and Spannagle, *The Complete Guide to Climate Change.*
§ Michael R. W. Rands et al., "Biodiversity conservation: challenges beyond 2010," *Science* 329, no. 5997 (2010).

mitigation by enhancing the biosequestration of carbon in plants and soils and thus reducing the accumulation of carbon in the atmosphere. Mitigation and adaptation strategies that make positive contributions to two or more interrelated issues also make sense in the context of limited resources and urgent time frames.

Land use change, especially tropical deforestation and habitat fragmentation, was the most important contributor to biodiversity loss during the last century and is expected to continue to be a major factor in the immediate future. Deforestation of intact rain forests of the tropics is by far the largest contributor to the approximately 20% of greenhouse gas emissions attributable to land use change since the industrial revolution and also a major cause of biodiversity loss. I return to an assessment of the pivotal role of tropical forests in Chapter 3.

Global warming and related climate changes are expected to accelerate biodiversity loss and to replace deforestation as the largest contributor to species loss during the twenty-first century. Habitat destruction, fragmentation, and pollution complicate and exacerbate changes attributable to warming temperatures. There are also rapidly emerging changes in major life cycle events of many species, known as changes in phenology, which are directly attributable to climate-induced changes in maximum and minimum air and water temperatures. Changes in the timing of spring greening and flowering, fish spawning, the emergence of insects, and migration dates are now well documented throughout the world. The geographic range of specific climatic conditions required by many species is changing at rates that make adaptation impossible and is thus contributing to high rates of extinction.[*] These climatic changes are compounded by habitat destruction, fragmentation, and barriers created by human infrastructure, such as highways and cities that block the migration of both plant and animal species.

A shift in the geographic distributions of both plant and animal species poleward in latitude and uphill in elevation has been documented as a strong pattern since around 1950. In a well-known study by Camille Parmesan and Gary Yohe, a database of more than 1,700 species with long-term observational records was utilized to analyze changes in phenology and geographic range. The post-1950 pattern analyzed by Parmesan and Yohe showed a worldwide poleward shift in species' distribution by an average of four miles per decade and a movement up mountainsides of 20 feet per decade. These movements were accompanied by an average change to earlier spring activity of 2.3 days per decade. These changes are much more rapid than the rates of change seen over geological time frames.[†]

---

[*] IPCC, *Climate Change 2001*, Third Assessment Report (Geneva: IPCC, 2001).
[†] Camille Parmesan and Gary Yohe, "A globally coherent fingerprint of climate change impacts across natural systems," *Nature* 421, no. 6918 (2003).

Research building on the pioneering research of Parmesan and Yohe shows that the changes they documented are accelerating. A 2011 meta-analysis of published research on the rates at which species have responded to recent warming found that the average rate of species movement away from the equator and to higher elevations has been two to three times faster than previously reported. The median distance in latitude change is now 16.9 kilometers (10.5 miles) per decade. This analysis also confirmed for the first time that the distances moved by species are greatest in areas with the highest levels of warming.[*]

Like many issues and findings related to climate change, this one is more complex than a simple linear relationship between temperature and species' geographical range. Water and energy availability, in addition to temperature, also constrain many species. Research recently conducted on a region of California that includes approximately one-half of the state and most of the mountain ranges north of 35 degrees latitude found widespread downhill shifts in the optimum elevations of many plants. The research confirmed an increase of mean annual temperature of about 0.6°C (1.08°F) in California between the 1930–1935 time period and 2000–2005. Precipitation in much of the study area also increased over the period and resulted in a decrease in "climatic water deficit" (a measure of the difference between potential evapotranspiration and precipitation). The plant species in the study area appear to be shifting to lower altitudes because of improved water availability despite an increase in temperature.[†] This research does not contradict the evidence of a widespread poleward migration of species or the evidence of many species moving to higher elevations as a response to warming. It serves as an illustration of the multiple interacting factors that must be considered in analyzing the complex ways that human actions affect ecological processes. This is the essence of ecological thinking, which seeks an understanding of nature as embodying multidimensional interactive relationships in the context of changing environments.

The current climate change-induced process of boreal forests invading tundra is another significant example of poleward migration and complex ecological interactions. As the high Arctic is warming at more than twice the global average, coniferous boreal forests are invading areas of tundra above the Arctic Circle from Siberia to Canada and Alaska. Trees are now growing in areas where they have not been seen for at least 1,000 years. Ecologists and climatologists are predicting that invading forests may alter the albedo, or reflectivity, of large areas across the Arctic. Most of the sunlight hitting snow and ice is reflected back into space, while most of the sunlight hitting

[*] I-Ching Chen et al., "Rapid range shifts of species associated with high levels of climate warming," *Science* 333, no. 6045 (2011).

[†] Shawn M. Crimmins et al., "Changes in climatic water balance drive downhill shifts in plant species' optimum elevations," *Science* 331, no. 6015 (2011).

open water or trees is absorbed and converted to heat. The primary tree species of the Arctic Ural Mountains, the Siberian Larch, has long grown in a stunted, shrubby form due to harsh conditions of cold, snow, and wind at the northern edge of its range. The spruce growing at polar tree lines in North America are similar in form and generally have been about 1 m high. Both the larch and the spruce are assuming upright form, developing seed, and spreading northward and to higher altitudes.[*]

Again, this process is not uniform or one directional. The advance of the tree line has been modest in Alaska and the changing growth rate of trees is variable, with some growing more slowly and others more quickly. As one would expect, another important variable is moisture. As the Arctic warms, much of it is expected to get dryer. In western North America there will likely be sites too dry to grow trees. Boreal forests hold large amounts carbon, mostly in their soils, and if the trees die, they decompose and CO_2 is released from the wood and from the unprotected soil. Fires are another outcome of reduced moisture. When the dense mats of organic matter in tundra are disturbed by fire, tree seeds are able to geminate at higher rates and forests migrate into previously treeless areas.[†]

An even more pervasive disturbance has been what ecologists call shrubbification. It begins with the arrival of small plants, often only about 3 inches tall at first. Spreading willows are common Arctic invaders. These plants capture snow, which insulates and warms the soil. The warmer soil activates bacteria that feed on and release to the atmosphere the carbon in the previously frozen soil. Satellite monitoring of the Alaskan Arctic shows a 39% increase of shrub cover since 1950 and an increase of biomass of 25–30%. The top meter of Arctic tundra stores up to 200 Gt (200 billion metric tons) of carbon. To put this in context, the Earth's atmosphere now holds about 825 Gt of carbon.[‡] Shrubs and trees sequester carbon, but boreal trees at the northern edge of their range grow so slowly that it is not likely that they will compensate for the carbon lost from the activation of warming soils. Trees are darker than the shrubs they replace and the snow they cover, so the net release of carbon is exacerbated by lower albedo (less reflectivity).[§]

The precise net impact of changes to the land carbon system is uncertain because of the complexity and variability of natural responses to human perturbations and because in many cases scientists simply have not yet collected and tested sufficient evidence to make specific predictions. Arctic warming by shrubbification, the loss of sea ice, and accelerated melting of the Greenland continental ice sheet have profoundly dangerous positive (warming) feedback potential. It is sobering to know that the projections

---

[*] Janet Raloff, "Forest invades tundra: ...And the new tenants could aggravate global warming," *Science News* 174, no. 1 (2008).
[†] Raloff, "Forest invades tundra."
[‡] Brian J. McPherson and E. T. Sundquist, *Carbon Sequestration and Its Role in the Global Carbon Cycle* (Washington, DC: American Geophysical Union, 2009).
[§] Raloff, "Forest invades tundra."

made by the IPCC of warming, sea level rise, ocean acidification, extreme weather events, and other climate change impacts do not generally include or account for the potentially devastating impacts of such powerful interactive feedbacks.* This is one reason the IPCC scenarios for the future likely underestimate the extent of future climate change. Much of the evidence published since the 2007 Fourth Assessment Report documents changes at or exceeding the upper ranges and rates projected by the IPCC.†

Another ecosystem change with complex causes and implications for biodiversity and warming is the degradation of grasslands. In North America at the time of European settlement there were approximately 360 million acres of short-grass, tall-grass, and mixed-grass prairie. There are now only about 70 million acres left. During the last thirty years 75% of grassland bird species have declined in abundance, more than any other group of North American birds. Small, ground-foraging species like the chestnut-collared longspur, large predators like the ferruginous hawk, and many others like the bobolink and greater prairie chicken are experiencing plummeting populations as complex grassland ecosystems disappear. Suburban sprawl, roads, power lines, oil wells, and wind turbines all have a negative affect on grassland species. Converting grasslands to croplands has an even larger impact. The rapid expansion of corn acreage to meet the demand for ethanol production in recent years has accelerated habitat loss. Corn is particularly damaging to grassland bird species because most of them will not nest in corn and conventional/industrial corn production uses high levels of chemical herbicides.‡

Since the Conservation Reserve Program (CRP) was established by the U.S. Department of Agriculture in 1985, landowners have been paid to put marginal cropland back into native vegetation. This program has provided critical wildlife habitat, improved water quality, and reduced soil erosion. At the end of 2010 about 31.4 million acres were enrolled in the CRP and almost one-third of those acres were in contracts set to expire during 2011. Much of that acreage was likely to be planted in corn, a conversion that will probably continue as long as federal agricultural and energy policies encourage and subsidize grain-based ethanol production. Current legislation mandates increases in corn ethanol production through 2015. The rapid increase in corn acreage contributes to the loss of both unbroken native prairie and the improved habitats on CRP lands that have been taken out of grain production and planted to grass and trees over the past twenty-five years. Prairie soils and land planted to native grasses and trees also lock up and hold much more carbon than tilled cropland. Plowing one acre of native prairie will release about half of its carbon, nearly 60 tons, over a fifty-year time period.§

---

* Dawson and Spannagle, *The Complete Guide to Climate Change.*
† Flannery, *Now or Never.*
‡ T. Edward Nickens, "Vanishing voices," *National Wildlife* Oct./Nov. (2010): pp. 23–29.
§ Nickens, "Vanishing voices."

Many of the ecological benefits realized because of the CRP are now in jeopardy due to current energy policy and the profit-driven promotion of corn ethanol in the United States. This serves as an instructive example of how government subsidies can promote ecologically sound land management practices or provide incentives for ecological degradation.

1.6.3 Ecological Change at the Poles

Let us go back to the Arctic to look at another set of ecosystem changes in the region that is experiencing perhaps the most extreme effects of human actions. From 1979 until 2010, Arctic sea ice cover declined an average of 11.5% per decade.[*] The U.S. Geological Survey (USGS) reported in 2007 that two-thirds of the Earth's 25,000 polar bears could disappear due to loss of sea ice habitat within fifty years if greenhouse gas emissions continue at current rates. This report was cited the following year when the U.S. Department of the Interior placed the polar bear on its list of threatened species. The polar bear is the most visible and celebrated of the species threatened by the loss of Arctic sea ice, but the ecosystem and food chain disturbances are wide and deep.

The semifrozen edge between floating ice and seawater is a surprisingly productive habitat that promotes the growth of microscopic plankton, which in turn serves as the base of the food chain. Ringed seals, the primary food source for most polar bears, eat fish from the rich and relatively warm waters beneath the ice. Other iconic species, including the walrus, white beluga whale, narwhal, and the huge bowhead whale, all live on, near, or under the Arctic sea ice. In addition to the ringed seal, bearded, spotted, ribbon, harp, and hooded seals all have a close connection to floating sea ice.[†] Killer whales have expanded their range as Arctic sea ice has thinned and declined in extent. They are now moving into the Hudson Bay, where the ice pack once kept them out. The killer whale threatens the rich marine mammal life in the Hudson Bay as the new top predator in an ecosystem previously dominated by the polar bear. Narwhals, the unique single-tusked whales whose numbers are rapidly declining, are under pressure from the killer whales. The narwhal likely adapted to life in the ice as a way to avoid killer whales, a successful adaptation now undermined by rapid climate change.[‡] All of these species and many others will have their ecological niches profoundly altered as the extent of sea ice declines and probably disappears entirely for much of the year and over much of the Arctic.

Similar changes related to declining sea ice and losses of the rich ecology of the ice edge are occurring in the Antarctic. The loss of krill (small

---

[*] Alexandra Witze, "Story one: swift action to cut greenhouse emissions could save polar bears: reductions could stabilize shrinking sea ice, study finds," *Science News* 179, no. 2 (2011).

[†] Alun M. Anderson, *After the Ice: Life, Death, and Geopolitics in the New Arctic*, 1st ed. (New York: Smithsonian Books, 2009).

[‡] Virginia Morell, "Killer whales earn their name," *Science* 331, no. 6015 (2011).

shrimplike crustaceans) has closely tracked the decline of Antarctic sea ice over the past fifty years. Krill feed on plankton and in turn serve as a primary food for penguins, albatrosses, seals, and whales. Antarctic sea ice was stable in extent from 1840 to 1950. Since 1950, ntarctic sea ice extent has declined by more than 20% and has been accompanied by a sharp decline in the krill population since 1976. The southwest Atlantic sector of the Southern Ocean is the heart of the krill population and is also the area of the most precipitous decline in both sea ice and krill numbers.[*] The land mass in this area is the northwestern Antarctic Peninsula, which has been warming rapidly. The peninsula is the location of the Palmer Research Station, one of the oldest research outposts in Antarctica, and the source of long-term data on temperature and ecological change. Average winter temperatures on the northwestern Antarctic Peninsula have risen 6.1°C (11°F) in the past sixty years and the average annual temperature has increased 2.8°C (5°F) since 1951. The warming has transformed all of the central features of the peninsula, including the glaciers that have receded 1,500 feet from the Palmer Station since the 1970s. Seabirds, most famously the Adelie Penguin, krill, and phytoplankton are all in decline along with the sea ice, which forms the foundation of the local ecology.[†]

In 2004 the Arctic Climate Impact Assessment report documented many changes affecting nearly every aspect of Arctic climate and ecology. A special focus of the report was an assessment of species of great importance to indigenous peoples of the Arctic. The caribou (or reindeer) is vital to indigenous people throughout North America and Eurasia. The numbers of Peary caribou on Greenland and Canada's Arctic islands has declined so rapidly since the 1960s that it was listed as endangered in 1991. It became of no use in the Inuit economy, as it could no longer be hunted. The Peary and other caribou are dependent on lichens for their winter food supply. With warming, much of the precipitation in the autumn months that once fell as snow now takes the form of rain, which freezes, covers the surface with ice, and makes the lichens inaccessible to caribou. A similar change has been documented in Finland. The long-term viability of the lichens themselves is also threatened as hungry caribou claw the plants from the soil, roots and all.[‡]

1.6.4 Vector-Borne Disease and Beetle Population Explosions

Not all species are harmed by a warming climate. Many insect species, like mosquitoes and ticks, often benefit from warmer temperatures. The geographical range, reproduction rate, and length of breeding season of many insects, including disease-carrying insects (vectors), are strongly affected by

---

[*] Flannery, *The Weather Makers*.
[†] Fen Montaigne, *Fraser's Penguins: Warning Signs from Antarctica*, 1st ed. (New York: Henry Holt, 2010).
[‡] Flannery, *The Weather Makers*.

temperature and moisture. Vector-borne diseases, including malaria, yellow fever, dengue fever, West Nile virus, Lyme disease, and encephalitis, are likely to accompany the insects that carry them to new and sometimes expanded areas. Climate change has resulted in an expansion of the range of vector-borne diseases to higher altitudes in Africa, Latin America, Asia, and Papua New Guinea. The mosquito that carries dengue and yellow fever, for example, has moved in recent years to altitudes that are about twice as high as elevations previously documented in Mexico and Columbia. Malaria is the most deadly of the vector-borne diseases, and its range is also expected to expand, especially in parts of southern and eastern Africa. Overall, Africa is expected to experience an expansion of malarial range over the next few decades, putting an additional 20–70 million people at risk.[*] The growth of human populations and the development of rapid global transportation systems "have made the world's biota more connected than at any time in Earth's history."[†]

The human-induced breakdown of biogeographic barriers results in the introduction of invasive species and pathogens at unprecedented rates. Many of the same pathogens and vector-borne diseases that thrive in warmer temperatures are also spread by travel and trade. The West Nile virus, HIV/AIDS, malaria, and dengue fever are examples of diseases that have spread rapidly via globalized transportation. Pathogens that affect wildlife and livestock, such as anthrax, rabies, and avian malaria, are spread by trade and travel, as are many diseases of wild plants and crops such as sudden oak death, potato blight, and chestnut blight. What all of these diseases and pathogens have in common is that their transmission is "linked with anthropogenic land use and increased abundance of domesticated animals and human-tolerant wildlife species."[‡]

Several species of bark beetle and the spruce budworm have been exceptionally successful in western North America during the warming period of the past thirty years or so. Bark beetle populations are normally limited by consecutive cold winters and cool summers. A combination of mild winters, warmer summer temperatures, and persistent drought have led to out-of-control bark beetle populations and the deaths of millions of trees from southern California and Arizona to the Yukon and Alaska. Starting in 1993 and continuing through 2010, more than 200 million Sitka, white, and Lutz spruce trees on more than 4 million acres have been killed in and near the Kenai National Wildlife Refuge on Alaska's south-central coast. This is the largest tree kill by insects in the history of North America.[§]

---

[*] Dawson and Spannagle, *The Complete Guide to Climate Change.*
[†] A. Marm Kilpatrick, "Globalization, land use, and the invasion of west Nile virus," *Science* 334, no. 6054 (2011): 323.
[‡] Kilpatrick, "Globalization, land use, and the invasion of west Nile virus," 326.
[§] Andrew Nikiforuk, *Empire of the Beetle: How Human Folly and a Tiny Bug Are Killing North America's Great Forests* (Vancouver: Greystone Books, 2010).

By 2003 spruce beetles had killed millions of trees on more than 300,000 acres in the Dixie National Forest near Salt Lake City, Utah. An estimated 100,000 trees are dying each day from beetle infestations in Colorado, Wyoming, and South Dakota. These and other areas of exceptional beetle infestation appear to be due primarily to a combination of drought-stressed tree populations and accelerated beetle reproduction cycles due to warmer temperatures.[*] Spruce budworm populations are also growing in the Arctic and sub-Arctic, as female budworms lay 50% more eggs at 77°F than at 59°F.[†] The death of trees has had a personal and visceral impact as I drive west of Denver, Colorado, on I-70 and watch the vast expanse of dead trees grow every year.

1.6.5 Biodiversity Loss

The anecdotes about several species provided above are offered in the hope that a few specific cases will serve to illustrate some of the variety and complexity of the interactive impacts that climate change is *now* having on life on Earth. There are a nearly inexhaustible number of other examples found in the scientific literature on biodiversity loss and its causes.[‡] The loss of biological diversity has much in common with climate change in terms of its roots in the nature and structure of industrial societies and also in terms of potential mitigation and adaptation strategies. This is a theme that will recur as we examine the carbon cycle, biosequestration, land management principles, and the policies and actions needed to address climate change and biodiversity loss. While anecdotes about a few species are instructive, interesting, and a cause for concern, the big picture on biodiversity loss in the context of climate change must be recognized as a profound threat to our collective survival. Only by having some understanding of the complexity and magnitude of this threat can we hope to develop effective mitigation and adaptation strategies. Like climate change, the loss of biological diversity is accelerating rather than abating and its primary causes and drivers are intensifying rather than diminishing.

The United Nations Convention on Biological Diversity (CBD) and the UNFCCC both emerged as a result of the Earth Summit held in Rio de Janeiro, Brazil, in 1992. In October 2010 the CBD held its 10th Conference of the Parties (COP 10) in Nagoya, Japan. The result was a series of decisions and publications regarding the status of global biological diversity and of efforts to slow the losses.

A report from COP 10, "Global Biodiversity Outlook 3: Executive Summary," provides a succinct summary of the status of global biodiversity eighteen years after the governments and nongovernmental organizations of

---

[*] Nikiforuk, *Empire of the Beetle.*
[†] Flannery, *The Weather Makers.*
[‡] Rands et al., "Biodiversity conservation: challenges beyond 2010."

the world officially recognized the problem. The review of evidence at COP 10 demonstrated that a top priority goal set in 2002, to significantly reduce the rate of biodiversity loss by 2010, had not been met. Biodiversity in all three primary components of genes, species, and ecosystems has continued to decline. Most species that have been found to be at risk of extinction are moving closer to extinction. Amphibians are at the greatest overall risk and coral species are deteriorating most rapidly in status. Nearly one-quarter of all plant species are estimated to be at risk of extinction. Vertebrate species declined in abundance by one-third from 1970 to 2006, with especially severe declines among freshwater species and in the tropics. Natural habitats are in decline in extent and integrity in most parts of the world, with freshwater wetlands, sea ice habitats, salt marshes, coral reefs, seagrass beds, and shellfish reefs showing especially severe declines. Fragmentation and degradation of forests, rivers, and other ecosystems have led to further loss of biodiversity and ecosystem services. Genetic diversity in crops and livestock is in continual decline.*

On the positive side, the evidence reviewed at COP 10 showed a significant slowing in the rate of loss for tropical forests and mangroves in some regions. This continues to be a priority for participating governments and represents an important step in an area of critical importance to biodiversity, the land-based carbon cycle, ocean shoreline protection, and global warming. Forest preservation, restoration, and creation projects from Indonesia to the Amazon, in Africa, and in nearly every part of the world will play a key role in any future success in mitigating and adapting to both biodiversity loss and climate change. I find it encouraging that the efforts devoted thus far to establishing, protecting, and restoring forests have had sufficient impact to draw notice as a bright spot in an otherwise bleak scientific review of biodiversity loss. Forests and their key role in climate change mitigation are a central focus of the carbon cycle and land management discussions in later parts of this book.

The principle "drivers" of biodiversity loss all result from human activities, and all of these drivers were determined at COP 10 to be either constant or increasing in intensity. The central drivers are habitat change, overexploitation, pollution, invasive alien species, and climate change. In clear and unequivocal language the authors of "Global Biodiversity Outlook" conclude that "the ecological footprint of humanity exceeds the biological capacity of the Earth" and the gap between the footprint and the capacity is getting wider. As the gap between the human footprint and the Earth's biological capacity widens, the essential "services" provided by ecosystems to humans are increasingly threatened. These services include the provision of food,

---

* Convention on Biological Diversity, "Global Biodiversity Outlook 3: Executive Summary," in *Conference of the Parties* (Nogoya, Japan: United Nations Environment Programme, 2010).

fiber, medicines, fresh water, pollination of crops, filtration of pollutants, and protection from natural disasters like droughts and floods.[*]

A group of scientists led by Michael R. W. Rands, of the Cambridge Conservation Initiative at Cambridge University, wrote the following abstract for a review article, "Biodiversity Conservation: Challenges Beyond 2010," published in *Science* in September 2010, shortly before the October 2010 meeting of the United Nations CBD:

> The continued growth of human populations and of per capita consumption have resulted in unsustainable exploitation of Earth's biological diversity, exacerbated by climate change, ocean acidification, and other anthropogenic environmental impacts. We argue that effective conservation of biodiversity is essential for human survival and the maintenance of ecosystem processes. Despite some conservation successes (especially at local scales) and increasing public and government interest in living sustainably, biodiversity continues to decline. Moving beyond 2010, successful conservation approaches need to be reinforced and adequately financed. In addition, however, more radical changes are required that recognize biodiversity as a global public good, that integrate biodiversity conservation into policies and decision frameworks for resource production and consumption, and that focus on wider institutional and societal changes to enable more effective implementation of policy.[†]

Human population growth and growth in per capita consumption are explicitly identified in this statement as resulting in an unsustainable exploitation of the Earth's biological resources that constitutes a threat to human survival. This bold, unequivocal declaration is supported by a detailed review of the essential services provided to human societies by biodiversity, defined as "the variety of genes, species, and ecosystems that constitute life on Earth," and a summary of its continued, accelerating decline. The decline in biological diversity has accelerated in most plant and animal groups despite an organized international effort for twenty years. The authors call for radical changes, including societal changes, to recognize biodiversity as a "global public good" and to integrate biodiversity conservation into policies and decisions regarding "production and consumption."[‡] These are revolutionary proposals from natural scientists that place the root cause of impending ecological collapse as residing in the nature and ecological impacts of industrial society.

---

[*] Convention on "Biological Diversity," Global Biodiversity Outlook 3.
[†] Rands et al., "Biodiversity conservation: challenges beyond 2010," 1298.
[‡] Rands et al., "Biodiversity conservation: challenges beyond 2010," 1298–1303.

1.7 Prelude to a Strategy

The idea that humanity has exceeded the biological capacity of the Earth is controversial in some quarters. There are reasoned arguments that perceived scarcities are not absolute and are caused primarily by inequality in a capitalist economic system designed to create profits and accumulate capital rather than to meet basic needs.* There is ample evidence that both absolute ecological scarcity and the nature of capitalist industrial economies underlie our ecological dilemma.† Ecologically sound approaches to reducing the severity of the unfolding global ecosystem collapse summarized here must address both the inequities and the pathological drive for unending growth that characterize contemporary industrial capitalism.‡

The material limits of the ecosphere require a transition from growth to forms of development that do not require ever greater throughputs of material resources and fossil energy. The urgent need for action on mitigation and adaptation strategies requires that we take concrete steps now. We do not have the luxury of waiting for current economic structures to change before acting. A structural transition away from fossil fuels to a noncarbon infrastructure will surely require profound changes in economic structures and capitalism.§ There are promising organized movements on every continent working in a variety of ways to initiate and further such a transformation.¶ Practical constraints require the use of existing and developing knowledge, both scientific and cultural, to implement workable strategies to mitigate and adapt to ongoing climate change and ecological degradation. We will surely modify strategies as we learn. We are dependent on the Earth's ecosystems for our material needs, yet it is possible to lose sight of this reality as it is mediated by complex social and technological systems. The implicit assumption that we have achieved a separation from nature, via technology, in industrial societies is a dangerous illusion.

It is not only our material needs that are met by the Earth's ecosystem services. Most of us also rely heavily on ecosystems for our aesthetic, spiritual, cultural, and psychological values and well-being. As a sociologist, I find it intriguing and ironic that natural scientists working on biodiversity loss can

---

* John Bellamy Foster, Brett Clark, and Richard York, *The Ecological Rift: Capitalism's War on the Earth* (New York: Monthly Review Press, 2010).
† Herman E. Daly, *Ecological Economics and Sustainable Development: Selected Essays of Herman Daly* (Cheltenham, UK: Edward Elgar, 2007); John Michael Greer, *The Wealth of Nature: Economics as if Survival Mattered* (Gabriola, BC, Canada: New Society Publishers, 2011); Richard Heinberg, *The End of Growth: Adapting to Our New Economic Reality* (Gabriola, BC, Canada: New Society Publishers, 2011); L. R. Brown, *World on the Edge* (New York: W. W. Norton, 2011).
‡ Heinberg, *The End of Growth*.
§ Larry Lohmann, "Capital and climate change," *Development and Change* 42, no. 2 (2011).
¶ Kolya Abramsky, *Sparking a Worldwide Energy Revolution: Social Struggles in the Transition to a Post-Petrol World* (Oakland, CA: AK Press, 2009).

explicitly acknowledge the importance of these nonmaterial realms while most social scientists seem to avoid any explicit reference to our collective dependence on ecosystems.* Ecological changes are rapidly undermining the habitability of the Earth. These changes cannot be halted before they radically transform the world of today's children. Ocean acidification and glacial melting, for example, have profound implications for global food security by 2050. These processes cannot be reversed in time to eliminate their impact on sea life and the availability of water for agricultural irrigation. We will need the participation of as many of us as possible to create viable collective strategies to adapt to the damage already done and to mitigate the intensity of additional damage.

Some are and will be effectively motivated to action by the impending loss of bird or mammal species that they find beautiful. Others find the respiratory and other illnesses caused by coal plant emissions unacceptable and will work on relevant legislation and regulation or the development of nonfossil energy sources. Still others are working to create new forms of perennial crops in order to slow the loss and degradation of soil and to establish a viable food system for the future. Some may love to nurture trees and participate in forest restoration projects. There is a great diversity of individual, organizational, corporate, and governmental actions needed. Nearly all of us can find ways to reduce our fossil energy use and level of material consumption. Learning ways to enhance and protect the carbon in soil and plants and keep it out of the overloaded atmosphere is a less obvious but essential component of any viable mitigation and adaptation strategy.

The examples above illustrate that human-induced ecological disruption and degradation are pervasive and transcend global warming. Many who are concerned with the future viability of human civilization and the ecological health of the Earth appear fixated on whether cumulative global warming has been 0.8°C or 1.0°C since a certain date, or whether atmospheric concentrations of CO_2 are 389 parts per million or 392 parts per million. The specific evidence and measurements are important, but less so than a recognition of the overall trends. This near fetish with slightly different numbers among many of those who accept the scientific consensus on global warming and its anthropogenic origins is similar, in its effect, to the deniers' unending emphasis on disputing the accuracy and variability of temperature measures by surface and satellite-based instruments. It draws attention and effort away from the broad body of evidence and a full recognition of the nature and extent of the dilemma confronting us. It also distracts attention and energy from the development and implementation of mitigation and adaptation strategies. The climate change deniers, like conservatives in the U.S. Congress, have largely succeeded in defining the terms of debate.[†] As suggested by Herman Daly, it is time to focus on first principles and to

---

* Foster et al., *The Ecological Rift.*
† Oreskes and Conway, *Merchants of Doubt*; Powell, *The Inquisition of Climate Science.*

take action. Careful and ongoing assessment of the nature of the problem can continue as strategies are implemented and refined.

———————

References

Abramsky, K. 2009. *Sparking a Worldwide Energy Revolution: Social Struggles in the Transition to a Post-Petrol World*. Oakland, CA: AK Press.

Anderson, A. M. 2009. *After the Ice: Life, Death, and Geopolitics in the New Arctic*, 1st ed. New York: Smithsonian Books.

Anderson, K., and A. Bows. 2008. "Reframing the climate change challenge in light of post-2000 emission trends." *Philosophical Transactions of the Royal Society A: Mathematical, Physical and Engineering Sciences* 366(1882):3863–3882.

Arndt, D. S., M. O. Baringer, and M. R. Johnson, eds. 2009. "State of the climate in 2009." *Bulletin of the American Meteorological Society* 91(7):S1–S224.

Blunden, J., D. S. Arndt, and M. O. Baringer, eds. 2010. "State of the climate in 2010." *Bulletin of the American Meteorological Society* 92(6):S1–S266.

Borenstein, S. 2011. "Skeptic finds he now agrees that global warming is real." *Lawrence Journal World*, October 31, 2011, 6A.

Brown, L. R. 2011. *World on the Edge*. New York: W. W. Norton.

Chen, I.-C., J. K. Hill, R. Ohlemüller, D. B. Roy, and C. D. Thomas. 2011. "Rapid range shifts of species associated with high levels of climate warming." *Science* 333(6045):1024–1026.

Committee on Stabilization Targets for Atmospheric Greenhouse Gas Concentrations. 2010. "Climate Stabilization Targets: Emissions, Concentrations, and Targets Over Decades to Millenia." http.www.nap.edu/catalog

Convention on Biological Diversity. 2010. "Global biodiversity outlook 3: Executive summary." In *Conference of the Parties*. Nogoya, Japan: United Nations Environment Programme.

Crimmins, S. M., S. Z. Dobrowski, J. A. Greenberg, J. T. Abatzoglou, and A. R. Mynsberge. 2011. "Changes in climatic water balance drive downhill shifts in plant species' optimum elevations." *Science* 331(6015):324–327.

Crutzen, P. J. 2002. "Geology of mankind." *Nature* 415(6867):23.

Daly, H. E. 2007. "Climate change: from 'know how' to 'do now'." CommonDreams.org., May 13, 2008, http://www.commondreams.org/archive/2008/05/13/8925

———. 2007. *Ecological Economics and Sustainable Development: Selected Essays of Herman Daly*. Cheltenham, UK: Edward Elgar.

Dawson, B., and M. Spannagle. 2009. *The Complete Guide to Climate Change*. New York: Routledge.

Dessler, A. E. 2010. "A determination of the cloud feedback from climate variations over the past decade." *Science* 330(6010):1523–1527.

Drake, N. 2011. "Environment: sulfur stalled surface temperature: coal emissions explain why warming stopped for a decade." *Science News* 180(3):17.

Edwards, P. N. 2010. *A Vast Machine: Computer Models, Climate Data, and the Politics of Global Warming*. Cambridge, MA: MIT Press.

Flannery, T. F. 2009. *Now or Never: Why We Must Act Now to End Climate Change and Create a Sustainable Future*, 1st ed. New York: Atlantic Monthly Press.
———. 2005. *The Weather Makers: How Man Is Changing the Climate and What It Means for Life on Earth*, 1st U.S. ed. New York: Atlantic Monthly Press.
Foster, J. B., B. Clark, and R. York. 2010. *The Ecological Rift: Capitalism's War on the Earth*. New York: Monthly Review Press.
Gillett, N. P., F. W. Zwiers, A. J. Weaver, and P. A. Stott. 2003. "Detection of human influence on sea-level pressure." *Nature* 422(6929):292–294.
Goddard Institute for Space Studies. 2008. "Global Temperature Trends: 2008 Summation." http://www.data.giss.nasa.gov/gisstemp/2008
———. "Research News." January 21, 2010. http:/giss.nasa.gov/research/news
Gramsci, A., and J. A. Buttigieg. 1992. *Prison Notebooks*. New York: Columbia University Press.
Greer, J. M. 2011. *The Wealth of Nature: Economics as if Survival Mattered*. Gabriola, BC, Canada: New Society Publishers.
Hansen, J. E. 2009. *Storms of My Grandchildren: The Truth About the Coming Climate Catastrophe and Our Last Chance to Save Humanity*, 1st U.S. ed. New York: Bloomsbury USA.
Heinberg, R. 2011. *The End of Growth: Adapting to Our New Economic Reality*. Gabriola, BC, Canada: New Society Publishers.
Hönisch, B., A. Ridgwell, D. N. Schmidt, E. Thomas, S. J. Gibbs, A. Sluijs, R. Zeebe, et al. 2012. "The geological record of ocean acidification." *Science* 335(6072):1058–1063.
Intergovernmental Panel on Climate Change. 2007. *Climate Change 2007: Synthesis Report*, Fourth Assessment Report (AR4). Geneva: IPCC.
———. 1997. "Guidelines for National Greenhouse Gas Inventories." http://www. ipcc.ch/pub/reports.htm
———. 2001. *Climate Change 2001*, Third Assessment Report. Geneva: IPCC.
Interim Secretariat for the UN Climate Change Convention and Secretariat of the UNEP/WMO Intergovernmental Panel on Climate Change. 2010. "United Nations Climate Change Bulletin." Châtelaine, Switzerland.
Jackson, W. 2010. *Consulting the Genius of the Place: An Ecological Approach to a New Agriculture*. Berkeley, CA: Counterpoint Press.
Kennedy, J. J. et al. 2010. "How do we know the world has warmed?" *Bulletin of the American Meteorological Society*, http:/www.ncdc.noaa.gov/bams-state-of-the-climate/2009.php
Kerr, R. A. 2011. "Vital details of global warming are eluding forecasters." *Science* 334(6053):173–174.
Kilpatrick, A. M. 2011. "Globalization, land use, and the invasion of West Nile Virus." *Science* 334(6054):323–327.
Kolbert, E. 2006. *Field Notes from a Catastrophe: Man, Nature, and Climate Change*, 1st U.S. ed. New York: Bloomsbury.
Lacis, A. A., G. A. Schmidt, D. Rind, and R. A. Ruedy. 2010. "Atmospheric CO_2: principal control knob governing Earth's temperature." *Science* 330(6002):356–359.
Lohmann, L. 2011. "Capital and climate change." *Development and Change* 42(2): 649–668.
Lynas, M. 2008. *Six Degrees: Our Future on a Hotter Planet*. Washington, DC: National Geographic.
McPherson, B. J., and E. T. Sundquist. 2009. *Carbon Sequestration and Its Role in the Global Carbon Cycle*. Washington, DC: American Geophysical Union.

Mills, C. W. 1960. *Images of Man: The Classic Tradition in Sociological Thinking*. New York: G. Braziller.

Montaigne, F. 2010. *Fraser's Penguins: Warning Signs from Antarctica*, 1st ed. New York: Henry Holt.

Morell, V. 2011. "Killer whales earn their name." *Science* 331(6015):274–276.

National Research Council. 2010. *Informing an Effective Response to Climate Change: America's Climate Choices*, ed. Board on Atmospheric Sciences and Climate. Washington, DC: National Academies Press.

Nickens, T. E. 2010. "Vanishing voices." *National Wildlife*: 22–29.

Nikiforuk, A. 2010. *Empire of the Beetle: How Human Folly and a Tiny Bug Are Killing North America's Great Forests*. Vancouver: Greystone Books.

Normile, D. 2010. "Counting the ocean's creatures, great and small." *Science* 330(6000):25.

Oreskes, N., and E. M. Conway. 2010. *Merchants of Doubt: How a Handful of Scientists Obscured the Truth on Issues from Tobacco Smoke to Global Warming*, 1st U.S. ed. New York: Bloomsbury Press.

Parmesan, C., and G. Yohe. 2003. "A globally coherent fingerprint of climate change impacts across natural systems." *Nature* 421(6918):37–42.

Perkins, S. 2009. "Earth: aerosols may have boosted carbon uptake: plant productivity could drop as skies continue to clear." *Science News* 175(11):14.

Powell, J. L. 2011. *The Inquisition of Climate Science*. New York: Columbia University Press.

Raloff, J. 2008. "Forest invades tundra: …and the new tenants could aggravate global warming." *Science News* 174(1):26–29.

———. 2012. "Insurance payouts point to climate change." *Science News*, Web edition, http://www.sciencenews.org/view/generic/id/337318/title/Insurance_payouts_point_to_climate_change

Rands, M. R. W., W. M. Adams, L. Bennun, S. H. M. Butchart, A. Clements, D. Coomes, A. Entwistle, et al. 2010. "Biodiversity conservation: challenges beyond 2010." *Science* 329(5997):1298–1303.

Richter-Menge, J., and J. E. Overland, eds. 2010. "Arctic Report Card 2010." http://www.arctic.noaa.gov/reportcard

Sabine, C. L., R. A. Feely, N. Gruber, R. M. Key, K. Lee, J. L. Bullister, R. Wanninkhof, et al. 2004. "The oceanic sink for anthropogenic CO_2." *Science* 305(5682):367–371.

Schellnhube, H., W. Cramer, N. Nakicenovic, T. Wigley, and G. Yohe, eds. 2006. *Avoiding Dangerous Climate Change*. Cambridge: Cambridge University Press.

Schmittner, A., N. M. Urban, J. D. Shakun, N. M. Mahowald, P. U. Clark, P. J. Bartlein, A. C. Mix, and A. Rosell-Melé. 2011. "Climate sensitivity estimated from temperature reconstructions of the last glacial maximum." *Science* 334(6061):1385–1388.

Scripps Institution of Oceanography. 2011. "Atmospheric CO_2." http://scrippsco2.uscd.edu/data/atmospheric_co2.html

United Nations Environment Programme. 2009. *Climate Change Science Compendium 2009*, ed. C. P. McMullen. Nairobi, Kenya: UNEP.

Vince, G. 2011. "An epoch debate." *Science* 334(6052):32–37.

Weart, S. R. 2008. *The Discovery of Global Warming*, rev. and expanded ed. Cambridge, MA: Harvard University Press.

Witze, A. 2010. "The final climate frontiers: scientists aim to improve and localize their predictions." *Science News* 178(12):24–28.

————. 2011. "Earth: 2010 ties record for warmest year: 20th century's global average exceeded for 34th year in row." *Science News* 179(4):17.

————. 2011. "Story one: swift action to cut greenhouse emissions could save polar bears: reductions could stabilize shrinking sea ice, study finds." *Science News* 179(2):5–6.

2

The Global Carbon Cycle and Terrestrial Biosequestration

2.1 Terrestrial Ecosystems and the Carbon Cycle Imbalance

2.1.1 Introduction

I have proposed that an important and underemphasized contribution to climate change mitigation and adaptation can be made by changing human use, care, and management of land and vegetation. The very same land management strategies that address climate change can also have positive effects on many of the other forms of ecological degradation that are endemic to industrial societies. This proposal is focused on terrestrial rather than marine ecosystems for several reasons. There are well-known and established methods to manage and safely influence the terrestrial carbon sink. Composting, planting trees and other perennial plants, keeping soil covered with mulch and preventing erosion by water and wind are examples. These methods can be implemented quickly on a small scale by individuals who want to take concrete, meaningful action. Collective actions on a larger scale can be implemented by local, state, regional, national, or international groups, governments, or corporations. Interventions in carbon sequestration processes in marine ecosystems or through carbon capture and storage (CCS) technologies are currently unproven, dangerous, and not available for near-term use. There is a rapidly developing body of evidence that enhanced sequestration and protection of carbon in the terrestrial system is a necessary and important component of a comprehensive strategy for stabilizing the global carbon cycle and reducing the degree and impacts of anthropogenic climate change and widespread ecological degradation.[*]

The global carbon cycle involves reservoirs or pools of carbon and the flux (movement) of carbon between these reservoirs. In simple terms, the carbon cycle imbalance created in the Anthropocene involves a massive and unnaturally rapid transfer of carbon from the terrestrial/biosphere

---

[*] Brian J. McPherson and E. T. Sundquist, *Carbon Sequestration and Its Role In the Global Carbon Cycle* (Washington, DC: American Geophysical Union, 2009).

(soil and plants) reservoir and the lithosphere (the Earth's crust, containing fossil fuels) reservoir to the atmosphere and ocean reservoirs. This transfer of carbon has been predominately accomplished by removing fossil fuels from the lithosphere, burning them, and transferring their carbon content in the form of carbon dioxide (CO_2) to the atmosphere and oceans. Much of the carbon stored in soils and plants has also been transferred to the atmosphere by burning, drying, and persistent and widespread disturbance and exposure of soil to sunlight and oxygen. In equally simple terms, this carbon cycle imbalance can be partially restored by leaving what remains of the fossil fuels in the lithosphere (or returning carbon to the lithosphere or deep oceans through unproven CCS technologies) and by restoring and protecting terrestrial ecosystems so they sequester and store more carbon for longer periods.

Improved land management practices are first and foremost a component of and a complement to other efforts to mitigate climate change by first stabilizing and then reducing atmospheric concentrations of greenhouse gases. A reduction to near zero, or at least an 80% reduction, in the emissions of CO_2 and the other pollutants generated by the burning of fossil fuels is now widely recognized by carbon cycle geoscientists as a necessary and essential component of any successful carbon cycle and climate stabilization strategy.[*] This is an appropriate and rational priority because the burning of fossil fuels has been and remains the primary source of anthropogenic greenhouse gas emissions, which are, in turn, the primary drivers of global warming, climate change, ocean acidification, and the underlying carbon cycle imbalance.

That portion of cumulative and ongoing anthropogenic greenhouse gases that are emitted by deforestation and soil degradation related to agriculture, forestry, and other forms of land use change must be addressed with strategies beyond reducing fossil fuel use. Although reducing fossil fuel use in agriculture and forestry may tend to improve soil quality and reduce soil degradation, the problems of agriculture and forestry, past and present, are greater than those created by fossil fuels alone. Changing land use and land management practices to reduce or reverse the emissions from deforestation and degradation of the terrestrial carbon sink are, I believe, underappreciated and underemphasized components of any comprehensive mitigation effort and also have a number of related and important ecological benefits.[†]

Agricultural and forest lands provide the elemental basis for most of the essentials of life. Without soil, civilizations perish, yet soil has been eroded

---

[*] Christopher B. Field and M. R. Raupach, *The Global Carbon Cycle: Integrating Humans, Climate, and the Natural World* (Washington, DC: Island Press, 2004); McPherson and Sundquist, *Carbon Sequestration and Its Role In the Global Carbon Cycle.*

[†] Wilfred M. Post et al., "Terrestrial biological carbon sequestration: science for enhancement and implementation," in *Carbon Sequestration and Its Role in the Global Carbon Cycle*, Geophysical Monograph Series 183 (Washington, DC: American Geophysical Union, 2009).

and degraded by human actions even before our ancestors first began the transition from hunting and gathering to planting crops at least 10,000 years ago.* When human populations were much smaller, the damage from perennial soil disturbance and deforestation was of local and regional concern. With human population exploding from less than 1 billion at the beginning of the Industrial Revolution to 7 billion in 2011, land use practices that once caused local degradation have become major contributors to the global carbon cycle imbalance. The economic growth and rising per capita consumption in the fossil-fueled industrial era have magnified the carbon and ecological impacts of the population explosion. The large and abundant Earth has become small and degraded.

2.1.2 Buying Time

Given our collective failure in recent years to make progress in reducing fossil fuel use and emissions, implementing strategies to increase and preserve terrestrial sequestration may be the best method available for "buying time" during the period required to transition the global energy system away from fossil fuels.† Research to develop and scale up CCS technologies, although expensive, energy intensive, and not ready for widespread use, will likely also continue. Despite its current limitations and lack of availability, CCS will likely be a necessary component of an overall global warming strategy to reduce greenhouse gas emissions by 80% below 1990 levels by 2050, the level of reductions projected by the Intergovernmental Panel on Climate Change (IPCC) that are required to avoid dangerous climate change.‡ In a comparative analysis of oceanic, geological, and biological sequestration versus fossil fuel emissions reductions, the potential utility of terrestrial biological sequestration is enhanced by the fact that it is relatively cheap, can to a great extent be implemented immediately, removes CO_2 directly from the air (as opposed to CCS, which is suitable only for point source emissions such as a coal-powered electric plants), and is potentially reversible.§

Illustrating the concept of reversibility is a useful way to introduce some of the central features of enhanced terrestrial carbon biosequestration and to gain a vision of some concrete possibilities. A reversible biosequestration project, for example, might be designed to plant or naturally regenerate trees on marginal and degraded crop and grazing lands for a period of

---

* Wes Jackson, *Consulting the Genius of the Place: An Ecological Approach to a New Agriculture* (Berkeley, CA: Counterpoint Press, 2010).
† Post et al., "Terrestrial biological carbon sequestration," Vincent Gitz et al., "Is there an optimal timing for sequestration to stabilize future climate?," in *Carbon Sequestration and Its Role in the Global Carbon Cycle*, ed. Brian J. McPherson and Eric T. Sundquist (Washington, DC: American Geophysical Union, 2009).
‡ James H. Williams et al., "The technology path to deep greenhouse gas emissions cuts by 2050: the pivotal role of electricity," *Science* 335, no. 6064 (2012).
§ Gitz et al., "Is there an optimal timing for sequestration?."

ten–eighty or more years. After ten–twenty years a portion of the forested land that was planted with fast-growing trees, like poplar or willow, for example, could be harvested for biofuels. This could replace some fossil fuels or ethanol from grain. If the biomass used as a fossil fuel substitute was partially preserved as biochar through pyrolysis, a low- or no-oxygen form of combustion, the biochar could be incorporated into soil to enhance long-term carbon sequestration, soil fertility, and water retention.[*] Slower-growing hardwood species could be utilized as they matured for building material and biofuels, and to add organic matter to the soil. Wood products utilized as building material or furniture sequester carbon for the life of the structures or products produced, which can be decades to centuries beyond the life span of the tree. Using appropriate and existing woodlot and forest management practices, the organic carbon in the soils would be substantially higher after trees are harvested than when the project started. This would result in the sequestration of more carbon in the soil, and the tree stands provide wildlife and understory plant habitat to enhance biodiversity. The additional organic matter provided by the roots, leaves, and other plants and organisms on the afforested land improves the fertility, water retention capacity, and productivity of the soil as it is planted back to crops or grass for grazing. Some of the afforested land could remain in trees indefinitely for ongoing selective harvesting or remain partly in trees as part of an agroforestry component, integrating crop production and trees. The species planted or allowed to regenerate from existing seeds or roots would vary by local ecological and economic conditions, as would the use of the trees and the crops eventually planted in the now improved soil.

Removing a large portion of degraded global crop and grazing land from agricultural production to store carbon, even in a time of increasing food insecurity, is viewed as a relatively effective and low-cost option by many geoscientists who cite an urgent need for strategies to stabilize the carbon cycle imbalance.[†] This is an indication of how much greater are the costs of business as usual and the many limitations of other sequestration options. CCS technologies are expected to be more expensive than enhancing terrestrial biosequestration, take much longer to implement, have greater human and environmental risks, and may cause damage that is not reversible for centuries to millennia.[‡] I will return to CCS, which involves capturing CO_2 at its source (e.g., a coal-fired electric plant) and injecting it into the ground or ocean, and its limitations. CCS, if it becomes ecologically and economically viable, is decades away from widespread commercial application.[§]

[*] J. Lehmann, "A handful of carbon," *Nature* 447, no. 7141 (2007).
[†] McPherson and Sundquist, *Carbon Sequestration and Its Role in the Global Carbon Cycle*; Post et al., "Terrestrial biological carbon sequestration."
[‡] Gitz et al., "Is there an optimal timing for sequestration?."
[§] McPherson and Sundquist, *Carbon Sequestration and Its Role in the Global Carbon Cycle*.

Making the necessary transition to a near-total replacement of fossil fuels to stabilize climate involves transforming the very nature of industrial civilization and is proving to take time (to say the least, since there has been minimal progress to date). Indeed, global CO_2 emissions set an all-time record in 2010, with a 6% increase over 2009 emission levels, and placing global emissions at higher levels than the worst-case projections of the IPCC.[*] Managing land to enhance terrestrial sequestration of carbon provides a way to reduce atmospheric CO_2 "within a few years and by a means less drastic than the shutdown of industrial civilization."[†]

2.1.3 Adapting

Improved land management practices that sequester and protect carbon have the additional benefit of providing ways to adapt to and reduce the damage of climate change impacts that have not been avoided and are occurring now. Farmers in Mali confronted with severe and persistent drought, Pakistanis experiencing frequent flooding, and California farmers with reduced access to summer irrigation water may eventually benefit from the successful reduction of atmospheric CO_2 concentrations (mitigation) to avoid even more damaging changes in the future, but they also need strategies to adapt to the changes that are already happening. The speed with which climate change is occurring no longer allows a single-minded focus on mitigation alone. Many of the climate changes now evident were not expected to become reality—even a few years ago—until at least several decades in the future. Some environmentalists and others concerned about climate change have considered efforts to adapt to climate change as a form of capitulation and a distraction from mitigation efforts. Placing a high value on mitigation to avoid severe and unmanageable changes in the future does not preclude the necessity of addressing and adapting to the very real changes that are occurring now.[‡]

Adapting to climate change through land management practices is being planned and implemented in the United States, Europe, and many other nations, cities, and rural areas.[§] Well-planned planting of trees and cover crops, and replacing annual plants with perennials, for example, can help local areas adapt by making them more ecologically resilient and resistant to drought and other forms of extreme weather associated with anthropogenic climate change. These examples also have the potential to enhance

---

[*] Seth Borenstein, "Biggest jump ever seen in global warming gases," *Lawrence Journal World*, November 4, 2011.
[†] Post et al., "Terrestrial biological carbon sequestration," 83.
[‡] Mark Hertsgaard, *Hot: Living Through the Next Fifty Years on Earth* (Boston: Houghton Mifflin Harcourt, 2011).
[§] National Research Council, *Informing an Effective Response to Climate Change: America's Climate Choices*, ed. Board on Atmospheric Sciences and Climate (Washington, DC: National Academies Press, 2010).

the biosequestration of carbon and thus serve as part of a mitigation strategy if implemented on a sufficient scale.[*] There is a great diversity of well-
established and emerging land management practices to help us mitigate
and adapt to climate change. These practices also have a positive effect
on biodiversity, soil fertility, and water quality. The principles underlying
these practices, how they can and are being implemented in a wide range
of circumstances, and their multiple ecological and social benefits appear to
get lost in most of the discourse on climate change. However, policies that
encourage or subsidize such practices and discourage practices that degrade
land systems are emerging at the local, state, and international levels.[†]

2.1.4 An Unplanned Global Geophysical Experiment

There may be no more profound support for chemist Paul Crutzen's 2002
declaration that we have exited the Holocene epoch and entered a new
geologic epoch of human domination—the Anthropocene—than the fact
that arguably the most universal and elemental natural cycle on Earth has
been "fundamentally altered by human activities."[‡] All life on the Earth
is dependent on the continuous cycling of carbon between and among
the atmosphere, oceans, animals, plants, soils, rocks, and sediments. This
natural cycle was first interrupted by humans thousands of years ago
when deforestation and other alterations of the landscape began to accelerate the CO_2 and methane (CH_4) flux between the land system and the
atmosphere. The perturbation of the carbon cycle from deforestation and
land clearing associated with hunting and gathering, and later farming,[§]
has now been dwarfed by the mining and burning of coal, oil, and natural gas.

In a classic and succinct expression of the nature of the carbon cycle interruption represented by the growing use of fossil fuels since the Industrial
Revolution, Revelle and Suess noted in 1957: "Within a few centuries we are
returning to the atmosphere and oceans the concentrated organic carbon
stored in sedimentary rocks over hundreds of millions of years," which they
characterized as a "large-scale geophysical experiment."[¶] Revelle and Suess
were some of the earliest geoscientists to recognize the potential implications

---

[*] Stephen Henry Schneider, *Climate Change Science and Policy* (Washington, DC: Island Press, 2010).

[†] Schneider, *Climate Change Science and Policy*.

[‡] Eric T. Sundquist et al., "An introduction to global carbon cycle management," in *Carbon Sequestration and Its Role in the Global Carbon Cycle*, ed. Brian J. McPherson and Eric T. Sundquist (Washington, DC: American Geophysical Union, 2009), 1.

[§] Alexandra Witze, "Environment: human climate meddling got start long before dawn of petroleum era: clearing forests released greenhouse gases by the gigaton," *Science News* 179, no. 9 (2011).

[¶] Roger Revelle and Hans E. Suess, "Carbon dioxide exchange between atmosphere and ocean and the question of an increase of atmospheric CO2 during the past decades," *Tellus* 9, no. 1 (1957), 19.

of this unprecedented human interruption of the global carbon cycle. Their early interest in accurately measuring and documenting the human-induced changes in the distribution of carbon among the major reservoirs has developed into an urgent interdisciplinary search among geoscientists for ways "to stabilize the global carbon cycle and protect the habitability of the earth."[*]

This urgency and concern for the very habitability of the planet is driven by recognition of the cumulative and profound extent to which human actions have altered the natural carbon cycle and the rapid and ongoing acceleration of carbon emissions in recent decades. Humans have been removing from the geologic reservoir and transferring to the atmosphere and oceans increasingly large quantities of the organic carbon (coal, oil, and natural gas) produced by photosynthesis over hundreds of millions of years. The global volume of CO_2 emissions from fossil fuels and cement rose 71% from 1970 to 2000. During the last ten years of that period, from 1990 through 1999, global CO_2 emissions from fossil fuels and cement increased at a rate of 1.3% per year. From 2000 through 2007 the rate of increase more than doubled in one decade, to 3.3% per year.[†] The global economic recession of 2008–2009 reduced global emissions temporarily, but global CO_2 emissions returned to record levels in 2010.

2.1.5 Rate of Change

The natural carbon absorption (sink) capacities of the land and ocean carbon reservoirs have been and continue to be greatly exceeded by the accelerating rate of emissions resulting from human land use change and the burning of fossil fuels. A 2009 summary of evidence published by the American Geophysical Union concluded that about 40%–45%, 220 gigatons out of 550 gigatons, of the cumulative carbon produced by the burning of fossil fuels, deforestation, and soil degradation since about 1850 now resides in the Earth's atmosphere.[‡] This means, remarkably, that more than half of cumulative industrial era anthropogenic carbon emissions from all sources have been removed from the atmosphere by the natural carbon cycle. Without the ecological services of resilient natural ecosystems our current climate and carbon cycle imbalance would be much more severe.

The Earth's terrestrial and ocean carbon sinks may have once been even more effective than our currently degraded ecosystems in removing anthropogenic carbon from the atmosphere. Researchers from the Federal Polytechnic School in Lausanne, Switzerland, reported in 2011 that human land use practices had added about 350 gigatons of carbon to the atmosphere

---

[*] Steven W. Running et al., "Next-generation terrestrial carbon monitoring," in *Carbon Sequestration and Its role in the Global Carbon Cycle*, ed. Brian J. McPherson and Eric T. Sundquist (Washington, DC: American Geophysical Union, 2009), 66.
[†] Running et al., "Next-generation terrestrial carbon monitoring," 66.
[‡] Sundquist et al., "An introduction to global carbon cycle management."

prior to 1850.[*] This estimate is derived from simulations of population growth, deforestation, and removal of vegetative cover for agricultural purposes. The 350 gigatons estimate is much larger than earlier estimates based on ice core data and other computer simulations. If accurate, this means that there were substantial cumulative preindustrial emissions. The preindustrial carbon was effectively removed from the atmosphere by terrestrial and marine ecosystems. Geologic evidence shows stable atmospheric concentrations of CO_2 during at least the 11,000 years before the industrial era, with variations of less than 20 parts per million during the entire period.[†]

The rate of carbon emissions is the key variable in determining the capacity of the natural carbon cycle to adapt. Rapidly expanding peatlands have been offered as one explanation of how the terrestrial system absorbed the preindustrial carbon emitted by deforestation and expanding agriculture.[‡] The oceans were also cooler and less acidic during the preindustrial period and thus were likely capable of absorbing much of the CO_2 emitted relatively slowly by preindustrial land use changes. If 350 gigatons of carbon were emitted over 10,000 years, the average emissions per century would be 3.5 gigatons, or 0.035 gigatons per year. The natural carbon cycle was apparently capable of sequestering this rate of anthropogenic carbon release. Compare the rate of emissions due to land use changes in the preindustrial period to the approximately 10 gigatons of carbon now emitted annually from industrial sources and land use changes,[§] and it is perhaps not surprising that the natural carbon sink capacities of the terrestrial ecosystems and oceans have been exceeded. The 10 gigatons of anthropogenic emissions in 2007 were more than 286 times greater than the approximate average of 0.035 gigatons per year of human-induced emissions in the preindustrial period. The emissions due to land use changes in the preindustrial period were likely greater near the end of the period than near the beginning as human population increased, so this estimate of annual rate is not perfect. The importance of the rate of change, however, is essential when thinking about the carbon cycle and ecological processes more generally. Rapid changes of great magnitude are those most likely to result in devastating ecological consequences.

The great resilience of the Earth's natural carbon cycle has limits that were overwhelmed by the rate and volume of accelerating global emissions from fossil fuels, in addition to the also accelerating rate of land use changes. In geologic time scales, natural changes in the carbon cycle have also caused changes in atmospheric concentrations of CO_2. The continental glaciations associated with the recurring ice ages during the past several hundred thousand years, for example, were associated with lower atmospheric concentrations of CO_2. There is simply no evidence, however, to support a contention

[*] Witze, "Environment: human climate meddling."
[†] McPherson and Sundquist, *Carbon Sequestration and Its Role in the Global Carbon Cycle.*
[‡] Witze, "Environment: human climate meddling."
[§] Sundquist et al., "An introduction to global carbon cycle management."

that "natural processes" can explain the current imbalance in the global carbon cycle.[*]

One of the great uncertainties in the scientific analysis of climate change is the potential future response of the terrestrial and ocean carbon reservoirs to ongoing warming and higher concentrations of atmospheric CO_2. It is certain that the absorptive capacities of the land system and the oceans themselves have been altered by higher global temperatures and higher atmospheric concentrations of CO_2. There is also evidence that the rate of carbon absorption by the oceans is declining and will decline further as ocean surface waters continue to warm and acidify.[†]

The rate and cumulative volume of future sequestration by the land system will be substantially determined by future land use practices and by climate change. The potential volume of carbon that can be biologically sequestered in the terrestrial system using available best practices and by utilizing new technologies is likely to be large enough to significantly reduce atmospheric concentrations of CO_2 in comparison to continuing current forestry, agricultural, and other land management practices.[‡] Future atmospheric CO_2 concentrations will depend "on the balance of human emissions, natural processes that remove excess carbon dioxide from the atmosphere, and the sensitivity of land and ocean reservoirs to climate change and land use."[§]

2.1.6 Carbon Reservoirs and Flux: From Stability to Imbalance

The lithosphere, or geologic reservoir, is by far the largest carbon pool in the global carbon cycle. It contains about 78 million gigatons of carbon, including about 6,000 gigatons of carbon in the form of fossil fuels. This is huge in comparison to the carbon stored in the terrestrial ecosystems of soils and plants, for example, which contain fewer than 4,000 gigatons of carbon. The geologic reservoir, however, is not naturally involved in the annual movement (flux) among the atmospheric, terrestrial, and ocean reservoirs. Weathering of rock is slow, and balanced by sedimentation of mineral carbon back into rock formations, and involves a relatively small annual flux globally. Similarly, volcanoes and other long-term forms of cycling through the Earth's crust involve a balanced flux of less than 0.1 gigatons annually. These natural geologic annual fluxes are small compared to the 120 gigatons of annual flux between terrestrial vegetation and soils and the atmosphere.

---

[*] Sundquist et al., "An introduction to global carbon cycle management."

[†] Richard A. Feely et al., "Present and future changes in seawater chemistry due to ocean acidification," in *Carbon Sequestration and Its Role in the Global Carbon Cycle*, ed. Brian J. McPherson and Eric T. Sundquist (Washington, DC: American Geophysical Union, 2009).

[‡] Post et al., "Terrestrial biological carbon sequestration."

[§] Committee on Stabilization Targets for Atmospheric Greenhouse Gas Concentrations, *Climate Stabilization Targets: Emissions, Concentrations, and Targets over Decades to Millenia* (Washington, DC: National Academies Press, 2011), 60.

The second largest flux in the natural carbon cycle is the approximately 70 gigatons of annual flux between the atmosphere and the oceans.[*] The oceans contain about 38,000–40,000 gigatons of carbon.[†] Although terrestrial ecosystems store much less carbon than the other nonatmospheric reservoirs, they have the highest natural rate of carbon exchange with the atmosphere.

When thinking about the carbon cycle it is helpful to remember the difference between the gas, CO_2, and the element, carbon (C). One metric ton of CO_2 contains 0.273 metric tons of carbon, so it takes 3.67 metric tons of CO_2 to equal 1 metric ton of carbon. (See Appendix A for this and other conversion units.) When the International Energy Agency reports annual emissions from fossil fuels, they refer to gigatons (1 billion metric tons) of CO_2 gas. When ecologists studying global forests summarized the most recent estimates of global carbon sinks and sources to arrive at a relatively conservative estimate of 4.1 gigatons of net carbon accumulating in the Earth's atmosphere per year, they were referring to carbon, not CO_2.[‡] The Soil Carbon Center at Kansas State University, Manhattan, Kansas, estimates the net annual addition of carbon to the atmosphere at 6.1 gigatons, with a possible range of 4.5–6.5 gigatons.[§] When discussing carbon sequestration, the unit of measure is carbon, and when considering atmospheric concentrations of greenhouse gases the unit of measure is usually CO_2 or carbon dioxide equivalent (CO_2-eq). The carbon element (C) of CO_2 in the atmosphere is fixed (sequestered) in plant tissue by photosynthesis in the form of organic carbon. Some of the carbon captured through photosynthesis (biosequestration) in plants on land is then transferred to soils in the form of soil organic carbon (SOC). The process continues and, through metabolism and decay, most soil organic carbon eventually returns to the atmosphere.

Atmospheric CO_2 is cycled through other forms of carbon in time periods ranging from seconds to thousands of years. Carbon from atmospheric CO_2 is converted rapidly by photosynthesis in plants to carbon in leaves, stems, and roots. However, most of the carbon taken in by plants is released immediately back into the atmosphere in the form of CO_2. The carbon released by plants as they photosynthesize and grow is termed autotrophic respiration. In the terrestrial system, some of the carbon taken in by plants is retained through photosynthesis and converted into leaves, stems, and roots, and is biologically sequestered for months, years, decades, centuries, and even thousands of years. The relative permanence of biological sequestration varies depending, in part, on the life span of the plant and how long it takes to decay after it dies. A portion of the carbon in plants is sequestered in the soil as carbon is exuded from the roots and as plant

---

[*] Sundquist et al., "An introduction to global carbon cycle management."

[†] Soil Carbon Center, "What is the Carbon Cycle?," http://soilcarboncenter.k-state.edu/carb-cycle.html

[‡] Yude Pan et al., "A large and persistent carbon sink in the world's forests," *Science* 333, no. 6045 (2011).

[§] Soil Carbon Center, "What is the Carbon Cycle?"

tissue decays in contact with the soil. Redwood and sequoia trees in the U.S. Pacific Northwest, for example, can live thousands of years and can take hundreds of years to decompose and release (through heterotrophic respiration) the carbon stored in their trunks and branches. Some of the deep root tissue may be sequestered for thousands of years after the death of the tree.

When plant matter is buried in soils or sediments it becomes a food source for microorganisms, which respire much of the carbon back to the atmosphere as CO_2 and also provide a supply of nutrients required by plants to grow and continue the photosynthesis that is the foundation of the carbon cycle. Some of the carbon converted to plant tissue is also taken up by plant-eating animals to produce the energy they need to survive; some is retained in their bodies, but most, as in plants and microorganisms, is respired back into the atmosphere as CO_2. Generally the deeper carbon is buried in the soil, the longer it will stay sequestered. Moisture levels, temperature, canopy cover, soil composition, and other factors affect the relative permanence of the carbon sequestered in terrestrial vegetation and soils. Microorganisms in the soil convert a portion of decaying plant and animal tissue into long-lasting SOC.

The central variables in terrestrial carbon sequestration are the total volume of carbon sequestered through photosynthesis (net primary production) and the length of time that it remains in plant tissues or soils (permanence). SOC is commonly divided into three forms or pools, depending on its permanence or residence time in soils. The active pool stores carbon from a few months to a few years and it is then respired or volatilized back into the atmosphere. This pool accounts for less than 5% of total SOC. The intermediate or slow pool stores carbon for decades and constitutes 20%–40% of the total. The longest-lasting, recalcitrant pool can store carbon, if undisturbed, for centuries to thousands of years and accounts for 60%–70% of total soil organic carbon.[*] The amount of soil organic carbon in soils is directly affected by the frequency of disturbance and the presence and extent of vegetative cover.[†]

The global store of carbon in soils, about 3,200 gigatons, is nearly five times greater than the carbon contained in vegetation, approximately 680 gigatons. The mass of carbon in soils and vegetation combined, 3,880 gigatons, is more than 6.5 times the carbon content of the preindustrial atmosphere, which was about 590 gigatons. The size of the carbon store in terrestrial soils and vegetation relative to that in the atmosphere has been significantly altered by human land use changes and fossil fuel emissions. The preindustrial ratio of terrestrial carbon to atmospheric carbon has been reduced from more than 6.5 times more carbon in terrestrial

---

[*] Soil Carbon Center, "What is carbon?," http://soilcarboncenter.k-state.edu/carbon.html
[†] R. F. Follett, "Soil management concepts and carbon sequestration in cropland soils," *Soil and Tillage Research* 61, no. 1–2 (2001).

systems than in the atmosphere to about 4.75 times. This change is the result of a 37%–40% increase in atmospheric carbon attributable to human activities.[*]

Human activities have altered each of the major carbon cycle processes. Forestry, agriculture, and other forms of land use have caused vast changes to the Earth's land cover. These activities have claimed and redirected one-quarter or possibly as much as 30% of global net primary production (NPP).[†] NPP is the net photosynthetic carbon uptake remaining after plants release through autotrophic respiration the CO_2 needed for their own metabolism. The approximately one-quarter of global NPP redirected by humans is used for the production of food (and feed for domesticated animals), clothing, shelter, and fuel.

Deforestation and soil degradation resulting from past and current land uses accelerate the release of CO_2 from dead plants and from disturbed and exposed soils. Enhanced rates of erosion and sedimentation accelerate both the exposure and release into the atmosphere of soil organic matter and its burial. Soil erosion involves one of the more contradictory and counterintuitive impacts of human actions; accelerated soil erosion appears to increase net carbon sequestration in some cases despite its many other negative impacts, including declines in soil productivity and sedimentation of lakes and streams. The erosion of soil carbon from uplands to lower-lying land, streams, and reservoirs often buries, and thus sequesters, carbon. This complicates the measurement and monitoring of terrestrial sequestration related to land management practices.[‡] The extensive soil erosion that has occurred on agricultural soils, however, creates much of the potential opportunity for enhancing future terrestrial sequestration of carbon by restoring the carbon to crop and pasture land that has been eroded and degraded by human use and abuse.

Research sponsored by the American Geophysical Union found that a net 1.5 gigatons of anthropogenic carbon emissions are sequestered annually by vegetation and soil and a net 2 net gigatons are absorbed annually by the oceans. Of the 120 gigatons of annual carbon exchange between terrestrial ecosystems and the atmosphere from primary production, about 3 gigatons remain in plants and soils and the remainder is respired back into the atmosphere. About half of this is currently lost due to land use practices that lead to deforestation and soil degradation. In the case of marine ecosystems, about 2 of the 70 gigatons of annual exchange remain in the ocean. There has been a recent increase in land-based sequestration that is due primarily to expanding temperate forests, especially in China, North America,

---

[*] Sundquist et al., "An introduction to global carbon cycle management."

[†] Post et al., "Terrestrial biological carbon sequestration."

[‡] Kristof Van Oost, Hendrick Van Hemelryck, and Jennifer W. Harden, "Erosion of soil organic carbon: implications for carbon sequestration," in *Carbon Sequestration and Its Role in the Global Carbon Cycle*, ed. Brian J. McPherson and Eric T. Sundquist (Washington, DC: American Geophysical Union, 2009).

and Europe, a decreasing rate of tropical deforestation,[*] and from the CO_2 fertilization effect, which involves increased photosynthesis due to higher atmospheric concentrations of CO_2.[†]

Carbon sequestration attributable to the CO_2 fertilization effect was once assumed to be large and was predicted by some to be the main driver of the future terrestrial carbon sink. Early climate models projected that enhanced photosynthesis from the CO_2 fertilization effect would be responsible for a cumulative land uptake of up to 600 gigatons of carbon during the next century in a scenario of doubled atmospheric CO_2. These projections are not supported by our recently improved understanding of the carbon cycle. The CO_2 fertilization effect is a small component of net sequestration.[‡]

The net exchange of 2 gigatons of carbon from the atmosphere to the oceans per year is the result of a differential between the pH (potential of hydrogen, degree of acidity) of the oceans and the increased concentration of CO_2 in the atmosphere. This increased exchange is making the oceans more acidic, which is reducing the pH differential between the oceans and the atmosphere and will reduce the rate of carbon absorption over time. Another climate change–carbon cycle feedback that affects the flux between the atmosphere and the oceans involves complex oceanic circulation changes due to warming and changes in atmospheric concentrations of CO_2. Climate change appears to be reducing global ocean carbon intake because stratification (less mixing) of surface water is increasing as circulation is slowing.[§]

The cumulative anthropogenic changes to the global carbon reservoirs since the Industrial Revolution include about 340 gigatons of carbon that have been emitted from the geologic reservoir in the form of burned fossil fuels. An additional 210 gigatons of carbon have been emitted from vegetation and soils during the same period due to deforestation and land degradation. Of these 550 gigatons of cumulative industrial era anthropogenic carbon emissions, about 180 gigatons have been sequestered in the terrestrial system and approximately 150 gigatons have been absorbed by the oceans. The remaining 220 gigatons have accumulated and remain in the atmosphere. This, in simple and direct terms, describes the nature and magnitude of the carbon cycle imbalance caused by human actions.[¶]

By burning fossil fuels and degrading the land system humans have released carbon from the geologic and terrestrial reservoirs at a rate greater than it could be sequestered by the terrestrial and ocean systems. The degree

---

[*] Sundquist et al., "An introduction to global carbon cycle management,"; Pan et al., "A large and persistent carbon sink in the world's forests."

[†] Rattan Lal, "Carbon sequestration," *Philosophical Transactions of the Royal Society B: Biological Sciences* 363, no. 1492 (2008).

[‡] Nicolas Gruber et al., "The vulnerability of the carbon cycle in the 21st century: an assessment of carbon–climate–human interactions," in *The Global Carbon Cycle: Integrating Humans, Climate, and the Natural World*, ed. Christopher B. Field and Michael R. Raupach (Washington, DC: Island Press, 2004).

[§] Gruber et al., "The vulnerability of the carbon cycle in the 21st century."

[¶] Sundquist et al., "An introduction to global carbon cycle management."

to which the natural sinks of the carbon cycle are being overwhelmed by anthropogenic emissions appears to be increasing in recent years. The approximately 60% of cumulative anthropogenic carbon emissions that has historically been absorbed by the terrestrial and ocean reservoirs has diminished in recent years to about 50%.[*]

The proportion of anthropogenic carbon emissions that will be absorbed by the oceans and terrestrial ecosystems with a continuation of current land use and management practices is expected to continue to decline as fossil fuel emissions increase, the oceans continue to warm and acidify, and degraded terrestrial ecosystems store less carbon. If we are to stabilize the global carbon cycle and protect the habitability of the earth, the options are limited and quite clear, and include reducing carbon emissions and increasing sequestration. The evidence is also clear that we need to do as much of both as quickly and safely as possible while also addressing the full range of related ecological threats summarized in Chapter 1. Enhancing biosequestration by changing land use and management practices is an essential component of any viable plan of action because it can be implemented rapidly, is relatively inexpensive, and has a range of associated ecological benefits.[†]

2.2 Enhanced Carbon Sequestration

The term "carbon sequestration" refers to processes that remove CO_2 from the atmosphere or capture it at emissions sources. There are two major forms of natural carbon sequestration in terrestrial ecosystems, termed biotic when referring to storage in plants and pedologic when referring to storage in soils. The other major natural form of sequestration involves absorption of carbon in the oceans. Photosynthesis by phytoplankton and other marine plants and uptake of CO_2 at the oceans' surface, as the slightly acidic gas bonds with carbonate ions, are the marine sequestration avenues.

2.2.1 CCS Technologies

Research to develop abiotic, engineered CCS technologies to enhance sequestration in the oceans and in the geologic reservoir is ongoing despite somewhat discouraging results to date. CCS technologies for carbon sequestration in both the oceans and the geologic reservoir may—if they can be

---

[*] Sundquist et al., "An introduction to global carbon cycle management."
[†] Field and Raupach, *The Global Carbon Cycle*; McPherson and Sundquist, *Carbon Sequestration and Its Role in the Global Carbon Cycle*.

successfully developed, scaled up, and safely implemented—be very useful in the long run to mitigate dangerously high temperatures. One of the attractions of abiotic sequestration technologies is the potential to store thousands of gigatons of carbon in the largest of the carbon reservoirs, the Earth's mineral crust and the deep oceans.[*] CCS is attractive to many, no doubt, because it is a centralized technology that has the potential to allow continued use of fossil fuels and thus avoid major changes in energy sources or rates of consumption. However, these methods are expensive, fossil fuel intensive, complex, and not yet available for widespread deployment. Oceanic sequestration strategies that involve pumping CO_2 into the deep ocean from concentrated sources like coal-fired power plants are constrained by risks of accelerated ocean acidification and related threats to marine ecosystems. CCS technologies for sequestration in geologic reservoirs are also complex, energy intensive, and plagued with substantial risks and uncertainties associated with carbon leakage and high costs.[†]

One large-scale, centralized technological method of sequestering carbon is known as scrubbing and mineral carbonization. This is a two-step process that involves scrubbing or chemical absorption of CO_2 using an amine or carbonate solvent and then heating the solvent to capture CO_2 in stable rock carbonates like limestone. Some or all of these technologies may prove useful in the future as they have the potential for sequestering large volumes of carbon. They are not available, however, to meet immediate needs for mitigation of the current and intensifying carbon and climate imbalance. They also do not have the ancillary ecological benefits, such as enhanced soil fertility and biological diversity related to biological sequestration in terrestrial systems. Abiotic technologies will also likely be limited to removing or preventing emissions from large sources like power plants and are not yet able to remove CO_2 directly from the atmosphere. It is likely that efforts to develop abiotic technologies to remove CO_2 directly from the air will also continue. The abiotic CCS technologies described here, if successfully developed, will likely not be available for deployment until after at least 2025.[‡] Perhaps the greatest danger associated with CCS technologies is a delay in implementing emissions reductions and biosequestration practices based on a faith in the future development of a technological fix.

2.2.2 Terrestrial Biosequestration

About half of the anthropogenic CO_2 released in the past few decades has remained in the atmosphere; about 30% has been sequestered by the oceans and 20% by terrestrial ecosystems.[§] The absorption of CO_2 by the oceans is

---

[*] Lal, "Carbon sequestration."
[†] Gitz et al., "Is there an optimal timing for sequestration to stabilize future climate?"
[‡] Lal, "Carbon sequestration."
[§] Feely et al., "Present and future changes in seawater chemistry due to ocean acidification."

declining and expected to continue to decline as surface waters acidify and warm. Oceanic carbon sequestration is also limited by the availability of iron (Fe) and several studies have assessed the potential to enhance biotic marine carbon sequestration by iron fertilization. Thus far the ecological risks associated with this strategy appear to outweigh the benefits of limited increases in photosynthetic activity.[*] Although research into ways of enhancing carbon sequestration in marine ecosystems will continue, there are no ecologically safe and effective methods currently available or on the immediate horizon.

Biosequestration in the terrestrial system, also known as biological or ecological sequestration, involves utilizing and learning to intelligently manage the largest of the natural annual fluxes, or movements, in the global carbon cycle. Terrestrial biosequestration includes the storage of carbon in both aboveground and belowground biomass and in soils as soil organic carbon and soil inorganic carbon. Biological sequestration, obviously, does not have unlimited potential in finite terrestrial ecosystems. Saturation can occur in terrestrial ecosystems when forests, grasslands, or other plant communities mature to a point that there is a near balance between carbon sequestration and release of carbon in the form of CO_2 through autotrophic (metabolism) and heterotrophic (decomposition) respiration.

The long history of anthropogenic deforestation and soil degradation has created an opportunity to draw substantial quantities of carbon from the atmosphere and back into the plants and soils of restored terrestrial ecosystems. This process of enhanced biosequestration can continue to expand for at least several decades, and probably throughout the twenty-first century.[†] The permanence of carbon sequestered in terrestrial ecosystems is at risk from both natural and anthropogenic disturbances—increased fire and drought in a warming climate, for example. Despite these limitations, which can be managed to some extent, proven methods to enhance terrestrial biosequestration may be the best set of tools we have to immediately mitigate the climate and carbon imbalance that we have created, adapt to the unavoidable changes that have begun, and help lessen the severity of related ecological crises.

It bears reiterating that enhancing natural terrestrial biosequestration provides an opportunity to directly address perhaps that most profound complication of the problem presented by the carbon cycle and climate imbalance, the need for immediate action. Wilfred M. Post, of the Oak Ridge National Laboratory, reminds us that "the atmospheric stabilization of carbon dioxide at any concentration requires that net emissions level off and eventually drop to near zero" and this "requires transformation of the energy systems worldwide, which will require many decades for development and

---

[*] Lal, "Carbon sequestration."

[†] Lal, "Managing soils and ecosystems for mitigating anthropogenic carbon emissions and advancing global food security," *BioScience* 60, no. 9 (2010); Rattan Lal, "Sequestering carbon in soils of agro-ecosystems," *Food Policy* 36, Supplement 1 (2011).

deployment."[*] This stark dilemma brings us back to the necessity of buying time with immediate mitigation, adapting to unavoidable climate changes, and doing so in ways that also safely address widespread ecological degradation. Deliberate enhancement of terrestrial biosequestration is widely identified by those who have evaluated carbon sequestration practices and technologies as the best currently available option for pursuing these interconnected goals. A large and obvious question is, how much difference can it make? The magnitude of the contribution to these goals that can be made depends on a full range of future policies, knowledge development, the scale and rate of land management changes, and the uncertain biological response of the terrestrial system to ongoing climate change.

Estimates of the potential magnitude of enhanced global carbon biosequestration range from a minimum 0.5 gigatons to more than 5.0 gigatons per year. The most conservative of these, the 0.5 gigaton estimate, would result in a potential cumulative enhanced sequestration of between 39.7 gigatons and 55.6 gigatons of carbon by 2100. These increases are calculated to be from reforestation and improved forest management, improved management of cropland, and improvements in pastureland. The carbon sequestration benefits are attributed to applying current best practices and are expected to diminish near the end of the twenty-first century if these practices are applied on nearly all available land.[†]

Pacala and Socolow estimated in 2004 that changing land use and improving land management could provide up to 2 gigatons of increased carbon sequestration per year compared to business as usual within fifty years. They proposed that gradually increasing carbon sequestration through land use and management changes could contribute up to 2 gigatons of the 7 gigatons per year in increased sequestration and/or reduced emissions required for stabilization of atmospheric CO_2 at about 500 parts per million within fifty years, compared to the then projected 700 parts per million by 2050 under a business as usual scenario.[‡]

Rattan Lal, of the Carbon Management and Sequestration Center at Ohio State University, has estimated that the world's "agroecosystems," which include croplands, grazing lands, and rangelands, have the potential to sequester an additional 1.2–3.1 gigatons of carbon per year with restorative land management practices.[§] This estimate is for agricultural soils only and does not include an additional 1.0–2.0 gigatons of annual carbon sequestration potential estimated for global forest biomass by the IPCC and others. When these estimates for soils and forests are combined they result in estimates that range from 0.5 gigatons up to 5.1 gigatons of enhanced carbon sequestration per year in the terrestrial system. Using the Kansas State Soil

---

[*] Post et al., "Terrestrial biological carbon sequestration," 74.
[†] Post et al., "Terrestrial biological carbon sequestration," 74.
[‡] S. Pacala and R. Socolow, "Stabilization wedges: solving the climate problem for the next 50 years with current technologies," *Science* 305, no. 5686 (2004).
[§] Lal, "Sequestering carbon in soils of agro-ecosystems."

Carbon Center estimate of 6.1 gigatons of annual net carbon accumulation in the atmosphere, achieving these levels of enhanced terrestrial carbon sequestration would reduce the annual accumulation of carbon in the atmosphere by 8%–84%. Using the lower end of the most recent assessment of the potential enhancement of agricultural soils by Lal of 1.2 gigatons and the lower estimate for forests of 1.0 gigaton, a total of 2.1 gigatons would reduce annual carbon accumulations in the atmosphere by 34%. This constitutes a significant potential contribution to stabilizing atmospheric concentrations of CO_2 that can be initiated relatively quickly, buying time for slower strategies like CCS and reducing fossil fuel use through conservation and increased energy efficiency and by expanding noncarbon and low-carbon energy sources.

The uncertainties about the potential global magnitude of enhanced biosequestration in the terrestrial system are due to the complexities of measures and estimates of maximum global biological potential, the amount of land and other resources available, and social and economic constraints that will affect the rate and extent of implementation. Much of this uncertainty is related to the statistical process of scaling up and extrapolating to regional and global levels information that is robust and directly measureable on a specific parcel of land. The evidence is clear that management options such as afforestation, reforestation, improved forest management, revegetation, reduced tillage, and a range of other practices increase carbon stocks in soils and vegetation.[*] These activities have carbon sequestration potentials ranging from about 100 kilograms per hectare per year to 1,000 kilograms per hectare per year (89–892 pounds per acre), depending on soil type, climate, and site-specific management.[†] Measurement of soil carbon content and biomass before and after implementing a specific practice is much simpler than extrapolating the results to diverse global ecosystems.

To facilitate policy development and implementation it is necessary to develop measures and monitoring, verification, and accounting systems to document changes in carbon stocks. If monetary incentives for carbon sequestration like those incorporated in the Kyoto Protocol, for example, are to be fully utilized to expand best practices, accounting and verification systems that apply to diverse ecosystems must be developed. Those measurement and accounting systems are now under development and use on regional, national, and global scales.[‡] It is not necessary to wait for sophisticated accounting systems to be perfected before learning about and implementing land use and management changes that have been proven to quickly and measurably enhance biosequestration and to have related ecological benefits.

---

[*] McPherson and Sundquist, *Carbon Sequestration and Its Role in the Global Carbon Cycle.*

[†] R. Lal et al., "Soil carbon sequestration to mitigate climate change and advance food security," *Soil Science* 172, no. 12 (2007).

[‡] McPherson and Sundquist, *Carbon Sequestration and Its Role in the Global Carbon Cycle.*

2.2.3 Overview of Established Land Use and Management Strategies

The estimates summarized previously of the potential global magnitude of enhanced annual carbon sequestration in terrestrial soils and vegetation are all derived from assessments of the potential associated with implementing established, proven land use and management methods. Common practices to enhance biomass in vegetation include afforestation, reforestation, forest management (e.g., thinning, pruning, removing invasive species, and fire suppression), revegetation of bare or degraded land, and managing cropland and grazing land to increase aboveground and belowground stocks of carbon.

Proven and established ways to enhance carbon stocks in soil also include afforestation, reforestation, improved forest management, revegetation, and the other means of increasing plant biomass. Any practice that increases net primary production (the photosynthetic product after autotrophic respiration) increases carbon in plants and also generally increases soil carbon stocks. Getting plant tissue into the soil requires leaving or placing leaves, stems, roots, and other organic matter in contact with the soil as the plant tissue decays. Increasing the volume of soil organic matter also requires protecting it from disturbance once it is sequestered.

There are many management strategies that improve the amount of carbon sequestered in soils and/or reduce the release of carbon from soils. A few important examples of such strategies for cropland include reduced or no tillage, mulching, conversion to deep-rooted crops, efficient use of animal manure and crop residue, conversion from annual to perennial crops, composting, improved crop rotations, use of cover crops and legumes, improved control of wind and water erosion, and conversion of cropland to grassland.*

Many of the practices used on cropland can also enhance carbon stores in soil on grazing land. Grazing land practices with carbon biosequestration potential include conversion to deep-rooted perennial species, utilizing a diversity of cool season and warm season forages, use of legumes, improved livestock management to minimize disturbance and maximize manure carbon returns, and managing stocking rates and timing to optimize growth and regrowth of forage. Effective and appropriate cropland and grazing land management strategies vary by soil characteristics, slope, annual precipitation, elevation, latitude, and management and land use history, among other factors, and must be selected with consideration of all of these factors. How these interrelated factors are considered in specific instances is illustrated when we consider some concrete examples in Chapter 4.

Forest biomes are the largest of the terrestrial reservoirs in both biomass and soil carbon storage. Planting trees in urban, suburban, and rural areas nearly always has a net beneficial effect on climate and carbon sequestration

---

* Pete Smith, "Engineered biological sinks on land," in *The Global Carbon Cycle: Integrating Humans, Climate, and the Natural World*, ed. Christopher B. Field and M. R. Raupach (Washington, DC: Island Press, 2004).

and often has other ecological benefits as well. Afforestation, especially when on a large scale, can also cause problems.

Afforestation involves planting trees on land that was not previously forested and involves some risks to water quantity and soil fertility in some areas.* Afforestation sites must be chosen carefully because trees generally reduce the albedo (reflectivity) of a tract of land. This albedo effect can partially offset the cooling (reduced radiative forcing) effect of carbon sequestration because trees absorb more solar energy and reflect less light than most surfaces (see the albedo table in Appendix B). Forest or plantation albedo is of special concern at high latitudes in areas with snow cover for extended periods each year. High-resolution satellite observations have shown, however, that reduced albedo does not entirely negate the cooling effects of afforestation. These observations have shown that afforestation between 40 degrees south latitude and 60 degrees north latitude always results in net cooling. Many of the areas with the highest net carbon drawdown (drawdown after albedo effects) are at the high latitudes and the overall efficiency of afforestation is about 50%. Afforestation efficiency refers to the net carbon drawdown divided by the total drawdown.† This means that albedo changes associated with afforestation reduce the net cooling effect of the carbon sequestered by about half, on average. These and similar findings also mean that measuring only the carbon sequestration of an afforestation project results in an overestimate of the net cooling effect of the carbon drawdown. As a result, several researchers have developed methods for incorporating changes in albedo when assessing the climate mitigation benefits of land use change projects and have proposed that these methods be incorporated into policy.‡

Land use and management strategies for increased carbon sequestration must be balanced with issues of stream flows, soil nutrients, biological diversity, albedo, and a full range of ecological and social considerations. This is another example of how holistic ecological thinking does not isolate one factor, outcome, or product from its context. These considerations are not so complex or difficult as to halt action; they do require the use of local knowledge resources.

In addition to forests, croplands, and grasslands, other important terrestrial carbon sinks to be protected, enhanced, or created include wetlands, urban forests and grasslands, inland aquatic systems, tundra, and taiga.§ Taiga, generally between 60 degrees north latitude and the Arctic Circle, provides an instructive example of how increased carbon biosequestration

---

* Robert B. Jackson et al., "Trading water for carbon with biological carbon sequestration," *Science* 310, no. 5756 (2005).
† Alvaro Montenegro et al., "The net carbon drawdown of small scale afforestation from satellite observations," *Global and Planetary Change* 69, no. 4 (2009).
‡ D. N. Bird et al., "Incorporating changes in albedo in estimating the climate mitigation benefits of land use change projects," *Biogeosciences Discussions* 5, no. 2 (2008); Montenegro et al., "The net carbon drawdown."
§ Smith, "Engineered biological sinks on land."

in the form of aboveground biomass does not always have a net mitigation effect on radiative forcing even though it may result in a small net increase in carbon biosequestration. Taiga refers to the transitional plant communities located between the polar tundra and the high-latitude boreal forests that generally consist of scattered trees and shrubs. Many areas of taiga and tundra now have trees and shrubs invading as the range of many species moves poleward with warming temperatures. This is a change that will likely contribute more to warming, by changing surface albedo, than it contributes to cooling through carbon sequestration from increased biomass. Increased tree cover in an area with extensive snow cover, like this transitional zone, reduces reflectivity and increases solar energy absorption. Encroaching trees and shrubs are on the edge of their range and tend to be stunted. The small amount of carbon that they incorporate in these harsh environments is likely more than offset by the reduced albedo of vegetation in a landscape with snow cover for much of the year. If the trees and shrubs absorb enough solar energy to cause melting of frozen soils, the carbon released from the enhanced microbial activity that results may also lead to carbon releases that exceed the carbon sequestered in the increased biomass.[*]

2.2.4 New and Developing Technologies for Enhanced Terrestrial Sequestration

In addition to the established land use and management practices just summarized (we return to these practices in much more detail later), there are also several new and developing technologies with the potential to enhance terrestrial sequestration including biotechnology, biomass carbonization, and deep-soil sequestration. Each of these methods is under consideration as a means to enhance terrestrial carbon sequestration up to several times greater than is possible with currently established methods.[†]

Given the growing influence of genetically modified organisms in contemporary industrial agriculture, it is not surprising that biotechnology is under intensive consideration as a means of enhancing carbon sequestration. Ongoing research and investment in plant and microbial biotechnology could lead to new ways to increase biomass production in agricultural crops and fast-growing trees. Areas of research include improvements in light-use efficiency and photosynthesis, improved root growth and acquisition of nutrients, and ways to overcome constraints on plant productivity posed by drought and temperature.[‡] Bioengineered increases in plant biomass could enhance the input of carbon in soils by increasing plant litter. Carbon sequestration could also possibly be enhanced by controlling the

---

[*] Janet Raloff, "Forest invades tundra: …And the new tenants could aggravate global warming," *Science News* 174, no. 1 (2008).
[†] Post et al., "Terrestrial biological carbon sequestration."
[‡] Post et al., "Terrestrial biological carbon sequestration."

chemical composition of plant litter to increase the rate and magnitude of litter sequestration in soils. The Rubisco enzyme in leaves plays a key role in photosynthesis and much of the research on the genetic engineering of plants to increase plant productivity has been focused on either increasing the amount of Rubisco in leaves or on changing the properties of the enzyme. Another area of inquiry involves attempts to alter the allocation of nutrient resources among various plant enzymes in an effort to increase carbon gain without an increase in the nitrogen required for metabolism.[*]

The potential for enhanced terrestrial carbon sequestration via plant biotechnology has been estimated to range from a cumulative increase of up to 130 gigatons of carbon by 2100 to a conservative estimate of at least 53 gigatons by 2100. These estimates are derived from an assumed increase of 2.5% and 1.0%, respectively, in global net primary productivity due to biotechnology. Global net terrestrial primary productivity is currently about 62 gigatons of carbon per year and about half is associated with managed forests and croplands. If biomass productivity in half of the managed forests and croplands could be enhanced by 10%, the amount of carbon available for sequestration annually would increase by about 1.5 gigatons. An assumption of a thirty-year adoption period and seventy years at full implementation produces the estimate of 130 gigatons by 2100. These projections are for bioengineered increases in biomass production only. Additional gains in carbon sequestration could result from modification in root turnover rates and from improved understanding of microbial communities that could lead to new management practices to maximize microbial conversion of residues to enhance carbon sequestration.[†]

There are significant risks associated with the use of biotechnology for enhanced carbon sequestration. Genetically modified organisms can escape and introduce engineered genes into wild populations. Engineered plants could also displace a range of other plants and reduce biodiversity. The risks associated with this emerging technology for carbon sequestration appear to be essentially the same as those associated with existing genetically modified crops. Although biotechnology has been widely adopted, particularly in maize (corn) and soybean crops, it has been strongly resisted in the European Union and Japan and among organic farmers. In the United States genetically modified (GM) crops have become so pervasive that it is impossible to completely exclude transgenic material from organic fields. About 93% of all soybeans grown in the United States, for example, are genetically modified. There is an ongoing dispute over who should bear the costs of preventing or slowing gene flow from GM crops to non-GM fields. The USDA continues to favor the GM industry and to approve new GM crops.[‡]

---

[*] Post et al., "Terrestrial biological carbon sequestration."
[†] Post et al., "Terrestrial biological carbon sequestration."
[‡] Erik Stokstad, "Can biotech and organic farmers get along?," *Science* 332, no. 6026 (2011).

Until the ecological risks associated with biotechnology are reduced and better understood, it seems reasonable and prudent to emphasize the widespread deployment of established and proven safe land use and management strategies. Similar to other high-risk engineering strategies, like releasing sulfur dioxide into the stratosphere to cause global dimming,[*] the use of GM organisms to enhance carbon sequestration may be useful sometime in the future to avoid or reduce catastrophic warming. Focused research should continue so that emergency measures are available if needed and their risks minimized.

If the accelerated rate of climate change impacts experienced during the past few years continues to exceed the scientific projections of the IPCC and many other climate scientists, high-risk strategies may soon be needed to avoid passing very dangerous thresholds. Tim Flannery concluded in an "assessment of probability" in 2009 that "there is now a better than even risk that, despite our best efforts, in the coming two to three decades Earth's climate system will pass the point of no return."[†] The point of no return referred to by Flannery was identified by James Hanson and his colleagues at the Goddard Institute of Space Studies as being that point at which the atmospheric greenhouse gas concentration reaches a level sufficient to cause catastrophic climate change ("a tipping point") and when that concentration has been in place long enough to initiate an irreversible process ("the point of no return"). Flannery argues that we have passed the tipping point but have a few years left before reaching the point of no return. He also states that to avoid reaching that threshold requires not only "a drastic change in energy use," but also "making full use of the tools we have at our disposal—and inventing new tools—to draw the pollution out of the air."[‡]

Biomass carbonization or pyrolysis is another tool beyond conventional land use and land management strategies that has the potential to greatly enhance terrestrial carbon sequestration. Pyrolysis involves carbonizing biomass by heating it in the absence or near absence of oxygen. The high carbon, charlike substance created by the incomplete combustion of biomass in low-temperature pyrolysis has the potential to sequester carbon for centuries to millennia when incorporated in soils. This is not an entirely new strategy. What has come to be known as biochar was first discovered in archeological studies of early human settlements and soils when partially burned biomass was found mixed in soils near human settlements in the Amazon region of South America. This charred biomass created enriched soils known as *terra preta*. *Terra preta* soil chars have been dated at 750–2,500 years old.[§]

Current methods of pyrolysis can also produce liquid and gaseous biofuels as well as biochar. Indeed, most recent research on pyrolysis has been focused

---

[*] Tim F. Flannery, *Now or Never: Why We Must Act Now to End Climate Change and Create a Sustainable Future*, 1st ed. (New York: Atlantic Monthly Press, 2009).
[†] Flannery, *Now or Never*, 100.
[‡] Flannery, *Now or Never*, 43.
[§] Post et al., "Terrestrial biological carbon sequestration."

on creating biochar as a supplemental product in bioenergy production.[*] A secondary focus has been on the creation of a cleaner-burning, low-cost, biomass-fueled stove technology. A low-emission, fuel-efficient stove that could be used for both cooking and heating and to also produce biochar has the potential to address several issues. A small household or community stove that produces biochar can be used to sequester carbon, improve soil fertility, and reduce air pollution and related health problems. Small-scale biochar stoves may prove especially valuable in much of the world where biofuels remain a principle household fuel source. The greatest potential for carbon sequestration using pyrolysis on a global scale is to incorporate the process into the growing biofuels industry. Pyrolysis can convert wood, grasses, crop residues, animal wastes, and even municipal wastes into biochar that has a carbon content, on average, about twice that of ordinary biomass.[†]

The amount of biochar and fuel produced by a given amount of biomass as well as the consistency and structure of the char depend on the biomass feedstocks, temperatures, and pressures used in the pyrolysis process. Chars are generally produced at temperatures ranging from 250°C to 800°C (482°F–1,472°F) and their porosity increases with temperature. All biochar is porous in structure. This is an important characteristic when it is incorporated into soil because the high surface area and pore structure provide a habitat for soil microorganisms that in turn help make soil nutrients like nitrogen and phosphorous more available to plants.[‡] In addition to its relatively high carbon content, the carbon in biochar is locked up in a much more recalcitrant form than the carbon in ordinary biomass. This durability is further enhanced when biochar is protected from oxidation and release into the atmosphere by incorporating it into soil. Like the carbon in ordinary biomass that is incorporated in soils, the relative permanence of the carbon in biochar will vary according to a number of factors, including the depth that it is buried, moisture conditions, temperature, and the pH and other properties of the soil. An estimate based on bituminous coal, which has a similar structure, places the half-life of char at 132 years.[§] Other estimates, based in part on the persistence of the chars dated in the *terra preta* soils of the Amazon region, range from hundreds to thousands of years. It is clear that the carbon in carbonized biomass is much more persistent than the majority of the carbon sequestered by the natural photosynthesis and decomposition process.

The overall contribution of biomass carbonization to global carbon sequestration is potentially large. Many scientific, economic, and policy factors will determine the actual amount of additional carbon sequestered during the coming decades using pyrolysis. The key technical variables include a sustainable level of biomass production, the efficiency of the carbonization

---

[*] Lehmann, "A handful of carbon."
[†] Lehmann, "A handful of carbon."
[‡] Jeff Schahczenski, *Biochar and Sustainable Agriculture* (Butte, MT: National Sustainable Agriculture Information Service, 2010).
[§] Post et al., "Terrestrial biological carbon sequestration."

technology employed, the amount of land available to store biochar and the durability or storage time of the carbon sequestered. To be economically viable, pyrolysis production facilities will require a continuous supply of locally available, inexpensive biomass feedstock.

Using a conservative estimate of a characteristic storage time of eighty years for biochar, biomass carbonization of 1% of global net primary production with a thirty-year startup period, Post et al. calculated that 19 gigatons of carbon could be sequestered in the next century. The fossil carbon emissions eliminated by the bioenergy produced from this same biomass more than doubles this amount.[*] Lehmann et al. estimated that replacing existing slash-and-burn agricultural practices with slash-and-char would reduce global anthropogenic carbon emissions from land use change by up to 12%, or 0.21 gigatons per year.[†]

When combined with bioenergy production, the production of biochar reduces carbon emissions by replacing fossil fuels and has been characterized as a way to make the bioenergy industry become "carbon negative."[‡] A carbon-negative industry or process is one that results in a net reduction of carbon in the atmosphere when subjected to a comprehensive, full life cycle analysis. This potential exists because pyrolysis can produce biofuels that displace fossil fuels and about half of the carbon in the biomass feedstock is carbonized and sequestered in biochar. This compares favorably with the approximately 10%–20% of biomass carbon sequestered after five–ten years of biological decomposition.[§] Incorporating biochar in soil is an interruption of the natural carbon cycle that slows the rate of carbon emissions from biomass by both increasing the fraction of carbon sequestered and slowing the rate of decomposition of the carbon captured by photosynthesis. Whether the pyrolysis process is or can be truly "carbon negative" when both biofuels and biochar are produced is an open question that is currently being examined. Experimental research on the residence time of biochar from varying feedstocks suggests that it may be possible to custom design biochar for specific soil types.

The production of biochar and liquid or gaseous fuels by carbonizing biomass with pyrolysis does not require expensive or time-consuming scientific or technological advances. The technology is simple, well understood, and can utilize a range of biomass feedstocks available in most regions of the earth. If ongoing research and demonstration projects show that it can be deployed on a large scale in an ecologically sound and economically viable manner, biomass carbonization may make a rapid and substantial contribution to enhancing terrestrial carbon sequestration. Widespread deployment would be more economically viable in the near

---

[*] Post et al., "Terrestrial biological carbon sequestration."
[†] J. Lehmann, J. Gaunt, and M. Rondon, "Bio-char sequestration in terrestrial ecosystems—a review," *Mitigation and Adaptation Strategies for Global Change* 11, no. 2 (2006).
[‡] Lehmann, "A handful of carbon."
[§] Lehmann et al., "Bio-char sequestration in terrestrial ecosystems."

term in the United States, for example, if the price of carbon on the Chicago Climate Exchange would rise significantly. Johannes Lehmann, of Cornell University (Ithaca, New York), estimated in 2007 that a carbon price of $37 per metric ton would make bioenergy production from pyrolysis economically attractive. Lehmann estimated that up to the equivalent of 30% of U.S. fossil fuel carbon emissions could be sequestered by pyrolysis using residue from forests used for timber production, fast-growing biomass on idle cropland, and crop residues from harvested cropland.[*] Removing crop residues for biomass energy production, however, risks depleting croplands of essential organic matter and fertility.

The conversion of biomass into biochar and its incorporation in soil are easily measured and monitored. The sources and amount of carbon in biochar can be identified by simple soil analysis. This makes accounting for the carbon sequestration from biochar much simpler than tracing soil carbon increases back to a specific agricultural practice or other photosynthetic source. The easy monitoring of carbon from biochar makes it potentially attractive for inclusion in the carbon emissions reduction programs under the Kyoto Protocol or other policies that encourage and subsidize terrestrial carbon sequestration.[†]

Unlike high-risk, complex, and expensive strategies like biotechnology, dimming the planet by polluting the stratosphere with sulfur dioxide, or CCS technologies, biomass carbonization does not entail significant ecological risks. Although not as simple as proven methods for increasing terrestrial carbon sinks like afforestation, reforestation, forest management, and improved management of cropland and grazing land, biomass carbonization may prove to be a valuable tool and an integral supplement to established practices. Carbon sequestered by biochar and incorporated in soils is not at risk of rapid or large-scale release. Biochar is much more stable and secure than the carbon sequestered in the aboveground biomass of a forest or grassland, which can be lost to the atmosphere very rapidly in a fire or because of land use changes. It also appears to be more stable than most ordinary sources of soil organic matter. Because of its multiple potential benefits and minimal risks, biomass carbonization seems to qualify as a "win–win" management option,[‡] one that effectively enhances carbon sequestration and has useful co-benefits. Biomass carbonization can produce biofuels that displace fossil fuels and biochar that sequesters carbon while improving soil fertility and water retention capacity. These results can lead to other related benefits such as lower fertilizer use and reduced pollution of surface and groundwater.[§]

---

[*] Lehmann, "A handful of carbon."
[†] Lehmann, "A handful of carbon."
[‡] Smith, "Engineered biological sinks on land."
[§] Schahczenski, *Biochar and Sustainable Agriculture.*

An estimated 3 billion people cook with biomass fuels, mostly using inefficient and highly polluting stoves and open fires. The pollution from inefficient and smoky burning of wood, coal, dung, crop residues, and a full range of other locally available biofuels causes extensive respiratory illness and environmental problems. The World Health Organization (WHO) estimates that 1.9 million people, mostly women and children, die each year due to exposure to smoke from biomass cooking. The World Bank reported in 2011 that 730 million tons of biomass are burned for cooking annually, resulting in 1.0 gigatons of CO_2 pollution. The World Bank is now supporting projects to assist local populations throughout the developing world to build or obtain cleaner, more efficient stoves.[*] Biomass burning for cooking and heating is also a significant source of black soot pollution that makes a significant contribution to radiative forcing by absorbing solar energy.[†] The International Biochar Initiative, World Stove, LLC, Cornell University, and a number of other organizations have been implementing and documenting demonstration and distribution programs of high-efficiency cook stoves that also produce biochar in regions that have concentrated use of biofuels for cooking. Demonstration projects that teach local people to manufacture and use biochar-producing cook stoves and the biochar produced from their use have been undertaken in several African nations, including Uganda, Kenya, Malawi, Zaire, Burkina Faso, Congo, and Niger. Projects have also been implemented in Haiti, the Philippines, and Indonesia.[‡] Instructions for making simple homemade biochar stoves out of widely available materials are now readily available on the Internet.

Biomass carbonization through pyrolysis offers a method of reducing one of the primary limitations of terrestrial biosequestration: the lack of permanence. The additional carbon accumulated in biomass and soil because of a land use or land management change can be lost to the atmosphere if the land use or management practice is reversed. All of the established methods of enhancing land carbon sinks—afforestation, reforestation, forest management, cropland management, grazing land management, and revegetation—are reversible and at potential risk of rapidly returning all or most of the carbon they sequester to the atmosphere. Carbon stored in soils and roots is more stable than that in aboveground biomass, but it too can be retuned to the atmosphere in a matter of years to decades, especially if the biomass covering it is lost to fire or other disturbance.

Biomass carbonization holds promise as an important tool for enhancing the volume and permanence of carbon in terrestrial ecosystems. Like other

---

[*] World Bank, "Cookstoves," http://www.climatechange.worldbank.org/content/cookstoves-report

[†] James E. Hansen, *Storms of My Grandchildren: The Truth About the Coming Climate Catastrophe and Our Last Chance to Save Humanity*, 1st U.S. ed. (New York: Bloomsbury USA, 2009).

[‡] Schahczenski, *Biochar and Sustainable Agriculture*; World Stove, "Biochar Cooking Stove Demonstration Projects," http://www.worldstove.com/about_2/pilot-programs/

strategies for enhancing terrestrial carbon sequestration, biochar technology has the potential to address multiple aspects of related ecological issues. When pyrolysis is used to produce liquid or gaseous biofuels it can replace some fossil fuels. When used for cooking, it can reduce air pollution and associated respiratory illness while conserving scarce biofuels. The biochar by-product improves soil fertility and water retention in addition to increasing the longevity of sequestered carbon.

A third developing technology for potentially enhancing established land use and management practices is deep-soil sequestration. It too is directed at overcoming the lack of permanence of the carbon sequestered in biomass and as organic matter in soil. Deep-soil sequestration is essentially an attempt to manipulate the depth that carbon penetrates soil through the addition of synthetic fertilizers and other soil amendments. Carbon stored deep in soils is much more stable than carbon stored closer to the surface and can persist for thousands of years under favorable conditions. At depths greater than 1 meter, soils have low oxygen levels and carbon is strongly stabilized on mineral surfaces. Both of these factors result in lower volatility and longer storage times. Naturally occurring deep soils also tend to have other characteristics that enhance the storage time of carbon; these characteristics include an acidic pH, abundant clay particles, and high iron oxide content.[*]

About one-third of the carbon stored in soils globally by natural processes is located at depths greater than 1 meter. It is estimated that deep soils have the global capacity to sequester about 165 gigatons of carbon for each meter of depth.[†] With this large potential capacity and the well-documented long storage times for deep-soil sequestration, scientists are exploring possible methods to enhance it. There are no other potential carbon sinks in terrestrial ecosystems that are comparable in magnitude to deep soils.

Fertilization methods commonly used in agriculture and silviculture may already be inadvertently enhancing the penetration of organic carbon in deep soils and are under examination as a way to systematically increase the vertical penetration of organic carbon. Nitrogen fertilizers, especially urea, interact strongly with soil organic materials and make humus more soluble so that it more readily penetrates deep soil horizons. Much of the commercial forest land in the southeastern United States is now routinely fertilized with nitrogen in the form of urea and with phosphate. It was discovered that there have been large increases of mobile dissolved organic carbon in these fertilized forest lands. When phosphate is used in conjunction with nitrogen it adds to the solubilization of soil organic matter. While these findings

---

[*] Post et al., "Terrestrial biological carbon sequestration."
[†] Ibid.

are promising, evidence is limited to needleleaf forests in the southeastern United States and to the deep soils (Andisol) found in that region.*

Careful analysis regarding the effectiveness of soil amendments to enhance the depth of soil organic carbon is needed. It is also important to examine the trade-off between deep soil carbon penetration and the availability of adequate nutrients closer to the surface to support vigorous plant growth. The amount of fossil fuel energy required to produce, transport, and apply the fertilizers must be compared to the amount of carbon sequestered. The greenhouse gases, including nitrous oxides, released in the manufacture and use of synthetic fertilizers under a variety of environmental conditions must also be considered.

Deep-soil sequestration may prove useful in addressing the carbon permanence problem. Like biotechnology and CCS technology, however, deep-soil sequestration is not ready for immediate application on a wide scale. Research will likely continue in part because these technologies hold profit potential for agribusiness in the cases of deep-soil sequestration and biotechnology, and for the coal industry and utilities in the case of CCS. Research should also continue in the interest of developing the most comprehensive portfolio of mitigation tools possible.

2.3 The Problem of Permanence

The lack of permanence of sequestered terrestrial carbon is a primary reason that the size of future carbon sinks in the terrestrial system is a central uncertainty in climate projections. Since the annual flux of carbon between terrestrial ecosystems and the atmosphere is the largest annual flux in the global carbon cycle, it holds great potential to have large positive or negative effects on atmospheric CO_2 concentrations. Warmer temperatures, the increasing frequency of wildfires, flooding, and drought all threaten the stability of terrestrial carbon.

The largest wildfire in the history of Arizona, and extensive fires in Texas, New Mexico, Oklahoma, and in the western part of my home state of Kansas during the spring and summer of 2011 brought home very clearly the profound fragility of terrestrial carbon stores. If enhanced carbon sequestration in terrestrial ecosystems is to make a significant and lasting contribution in a comprehensive strategy to stabilize atmospheric CO_2 and reduce radiative forcing, it must be done in ways that enhance permanence as much as possible. This means emphasizing and prioritizing practices that (1) enhance belowground biomass and soil carbon, (2) sequester carbon in wetlands and other moist ecosystems in areas that are less subject to disturbance from fire

---

* Post et al., "Terrestrial biological carbon sequestration."

and drought, (3) directly contribute to replacing and/or reducing fossil fuel consumption, and (4) increase the drought resistance of ecosystems.

These priorities suggest that afforestation, reforestation, deforestation avoidance, and improved forest management projects should include appropriate fire protection and management at their core. Long-term carbon storage is more secure in mature stands of diverse species, for example, than in single-species, same-age plantations. Selective harvesting of trees to improve the age and species diversity of a stand preserves resistance and resilience to fire, as well as disease and insect damage. Selective harvesting of trees also protects belowground biomass and soil organic carbon from exposure and drying due to loss of canopy cover. Using a portion of pruned branches and tops from harvested trees to locally produce biomass fuel to displace fossil fuels improves the carbon balance of forests planted or managed to reduce radiative forcing. Using wood to replace plastic and other fossil fuel–based materials further reduces fossil fuel consumption. In grasslands, planting or nurturing diverse native perennial species as opposed to monocultures of nonnative introduced species has similar impacts. Deep-rooted perennial species store carbon deep in the soil where permanence is enhanced.

The future size of the terrestrial land sink is strongly dependent on human land use and management, and in particular the actions taken to enhance the quantity and longevity of carbon in plants and soils. To fully appreciate the extent to which human actions will determine the size of the land carbon sink, it is useful to again compare scientific estimates of global biological potentials for sequestration with those potentials limited by economic and social constraints. In 2000 the IPCC estimated that total enhanced terrestrial carbon sequestration potential was about 2.5 gigatons per year. This estimate included all carbon sinks from the management of agricultural lands, grasslands, rangelands, and forests. It also included the protection and creation of wetlands and urban forests and grasslands, the management of deserts and degraded lands, and the preservation of sediments, aquatic systems, tundra, and taiga. This estimate of 2.5 gigatons of annual carbon sequestration included what IPCC scientists concluded were reasonable estimates of future human actions. It has been estimated that the biological potential, without consideration of social and economic constraints, for enhanced carbon sequestration in all of the terrestrial ecosystems is 5.65–8.71 gigatons of carbon per year.[*] We saw earlier that the total net carbon accumulating in the Earth's atmosphere due to human activities is 4–6 gigatons per year. The IPCC estimate equals about half of the annual net anthropogenic carbon accumulation in the atmosphere, and the estimate of biological potential with no social and economic constraints substantially exceeds the size of the net annual anthropogenic carbon imbalance. The potential of enhanced terrestrial sequestration is clearly substantial in relation to the global carbon imbalance.

---

[*] Smith, "Engineered biological sinks on land."

Given the emerging evidence of rapid climate change, it is prudent to examine every possible means to reduce atmospheric concentrations of CO_2, other greenhouse gases, and all anthropogenic sources of radiative forcing. As Tim Flannery and James Hansen have made clear, we have entered a period of such great risk of catastrophic climate change that we must develop a knowledge of, and be prepared to employ, even high-risk emergency measures. This includes being prepared to intentionally pollute the stratosphere with sulfur dioxide to deflect solar energy. Flannery also argues that the practical way to avoid such dangerous emergency measures is for the developed world (mostly the United States, Europe, and Japan, who created the climate dilemma) to finance a "clean development" mechanism for reducing carbon pollution in developing nations like India and China and at the same time rapidly implement land use and management changes in agriculture and forestry to reduce emissions and increase sequestration.[*]

In industrial capitalist societies, a disproportionate amount of resources, both intellectual and material, is often devoted to the development of new technologies to reduce labor requirements and improve profitability.[†] It is cruelly ironic that coal companies and utilities have failed to invest sufficient resources to develop CCS technologies to ensure their own future viability. As long as strong regulatory standards and enforcement are absent, including a direct price on carbon, we will not likely see CCS technologies in widespread use on electric power plants or other industrial sources of large-scale, concentrated emissions. Under the most optimistic of scenarios, CCS technologies will not be available for widespread use for decades at best, and serious questions about their safety remain.

Land use and land management practices that apply existing knowledge and technology are underutilized. The skewed allocation of resources toward expensive and unproven industrial technologies, which themselves require large fossil fuel inputs, detracts from the recognition and deployment of currently available, ecologically beneficial, and relatively inexpensive means of enhancing biological carbon sequestration. A widespread, rapid, and ecologically informed implementation of afforestation, reforestation, avoided deforestation, improved management of crop and grazing land, and careful preservation of existing wetlands, peatlands, and other natural carbon sinks will not solve the climate and ecological crises we face. These actions do have the potential to be important components— along with the most rapid reductions possible in the use of fossil fuels—of a multifaceted strategy to slow the ongoing damage to the global ecosphere. Nurturing the health of terrestrial ecosystems can also help people and many other species adapt to dangerous climate-related ecological degradation and changes.

---

[*] Hansen, *Storms of My Grandchildren*.

[†] John Bellamy Foster, Brett Clark, and Richard York, *The Ecological Rift: Capitalism's War on the Earth* (New York: Monthly Review Press, 2010).

References

Bird, D. N., M. Kunda, A. Mayer, B. Schlamadinger, L. Canella, and M. Johnston. 2008. "Incorporating changes in albedo in estimating the climate mitigation benefits of land use change projects." *Biogeosciences Discussions* 5(2):1511–1543.

Borenstein, S. 2011. "Biggest jump ever seen in global warming gases." *Lawrence Journal World*, November 4, 2011, 7A.

Committee on Stabilization Targets for Atmospheric Greenhouse Gas Concentrations. 2010. "Climate Stabilization Targets: Emissions, Concentrations, and Targets over Decades to Millenia." http://www.nap.edu/catalog

Feely, R. A., J. Orr, V. J. Fabry, J. A. Kleypas, C. L. Sabine, and C. Langdon. 2009. "Present and future changes in seawater chemistry due to ocean acidification." In *Carbon Sequestration and Its Role in the Global Carbon Cycle*, ed. B. J. McPherson and E. T. Sundquist, 175–188. Washington, DC: American Geophysical Union.

Field, C. B., and M. R. Raupach. 2004. *The Global Carbon Cycle: Integrating Humans, Climate, and the Natural World*. Washington, DC: Island Press.

Flannery, T. F. 2009. *Now or Never: Why We Must Act Now to End Climate Change and Create a Sustainable Future*, 1st ed. New York: Atlantic Monthly Press.

Follett, R. F. 2001. "Soil management concepts and carbon sequestration in cropland soils." *Soil and Tillage Research* 61(1–2):77–92.

Foster, J. B., B. Clark, and R. York. 2010. *The Ecological Rift: Capitalism's War on the Earth*. New York: Monthly Review Press.

Gitz, V., P. Ambrosi, B. Mangne, and P. Ciais. 2009. "Is there an optimal timing for sequestration to stabilize future climate?" In *Carbon Sequestration and Its Role in the Global Carbon Cycle*, ed. B. J. McPherson and E. T. Sundquist, 161–174. Washington, DC: American Geophysical Union.

Gruber, N., P. Friedlingstein, C. B. Field, R. Valentini, M. Heimann, J. E. Richey, P. R. Lankao, E.-D. Schulze, and C.-T. A. Chen. 2004. "The vulnerability of the carbon cycle in the 21st century: an assessment of carbon-climate-human interactions." In *The Global Carbon Cycle: Integrating Humans, Climate, and the Natural World*, ed. C. B. Field and M. R. Raupach. Washington, DC: Island Press.

Hansen, J. E. 2009. *Storms of My Grandchildren: The Truth About the Coming Climate Catastrophe and Our Last Chance to Save Humanity*, 1st U.S. ed. New York: Bloomsbury USA.

Hertsgaard, M. 2011. *Hot: Living Through the Next Fifty Years on Earth*. Boston: Houghton Mifflin Harcourt.

Jackson, R. B., E. G. Jobbágy, R. Avissar, S. B. Roy, D. J. Barrett, C. W. Cook, K. A. Farley, D. C. le Maitre, B. A. McCarl, and B. C. Murray. 2005. "Trading water for carbon with biological carbon sequestration." *Science* 310(5756):1944–1947.

Jackson, W. 2010. *Consulting the Genius of the Place: An Ecological Approach to a New Agriculture*. Berkeley, CA: Counterpoint Press.

Lal, R.. 2008. "Carbon sequestration." *Philosophical Transactions of the Royal Society B: Biological Sciences* 363(1492):815–830.

———. 2010. "Managing soils and ecosystems for mitigating anthropogenic carbon emissions and advancing global food security." *BioScience* 60(9):708–721.

———. 2011. "Sequestering carbon in soils of agro-ecosystems." *Food Policy* 36(Suppl. 1):S33–S39.

Lal, R., R. F. Follett, B. A. Stewart, and J. M. Kimble. 2007. "Soil carbon sequestration to mitigate climate change and advance food security." *Soil Science* 172(12):943–956.

Lehmann, J. 2007. "A handful of carbon." *Nature* 447(7141):143–144.

Lehmann, J., J. Gaunt, and M. Rondon. 2006. "Bio-char sequestration in terrestrial ecosystems—a review." *Mitigation and Adaptation Strategies for Global Change* 11(2):395–419.

McPherson, B. J., and E. T. Sundquist. 2009. *Carbon Sequestration and Its Role in the Global Carbon Cycle.* Washington, DC: American Geophysical Union.

Montenegro, A., M. Eby, Q. Mu, M. Mulligan, A. J. Weaver, E. C. Wiebe, and M. Zhao. 2009. "The net carbon drawdown of small scale afforestation from satellite observations." *Global and Planetary Change* 69(4):195–204.

National Research Council. 2010. *Informing an Effective Response to Climate Change: America's Climate Choices,* ed. Board on Atmospheric Sciences and Climate. Washington, DC: National Academies Press.

Pacala, S., and R. Socolow. 2004. "Stabilization wedges: solving the climate problem for the next 50 years with current technologies." *Science* 305(5686):968–972.

Pan, Y., R. A. Birdsey, J. Fang, R. Houghton, P. E. Kauppi, W. A. Kurz, O. L. Phillips, et al. 2011. "A large and persistent carbon sink in the world's forests." *Science* 333(6045):988–993.

Post, W. M., J. Amonette, R. A. Birdsey, C. T. Garten, Jr., R. L. Graham, Dr. R. C. Izaurralde, P. M. Jardine, et al. 2009. "Terrestrial biological carbon sequestration: science for enhancement and implementation." In *Carbon Sequestration and Its Role in the Global Carbon Cycle,* Geophysical Monograph Series 183. Washington, DC: American Geophysical Union.

Raloff, J. 2008. "Forest invades tundra: …and the new tenants could aggravate global warming." *Science News* 174(1):26–29.

Revelle, R., and H. E. Suess. 1957. "Carbon dioxide exchange between atmosphere and ocean and the question of an increase of atmospheric CO_2 during the past decades." *Tellus* 9(1):18–27.

Running, S. W., R. R. Nemani, J. R. G. Townshend, and D. D. Baldocchi. 2009. "Next-generation terrestrial carbon monitoring." In *Carbon Sequestration and Its Role in the Global Carbon Cycle,* ed. B. J. McPherson and E. T. Sundquist. Washington, DC: American Geophysical Union.

Schahczenski, J. 2010. *Biochar and Sustainable Agriculture.* Butte, MT: National Sustainable Agriculture Information Service.

Schneider, S. H. 2010. *Climate Change Science and Policy.* Washington, DC: Island Press.

Smith, P. 2004. "Engineered biological sinks on land." In *The Global Carbon Cycle: Integrating Humans, Climate, and the Natural World,* ed. C. B. Field and M. R. Raupach. Washington, DC: Island Press.

Soil Carbon Center. 2010. "What Is Carbon?" http://soilcarboncenter.k-state.edu/carbon.html

———. 2010. "What Is the Carbon Cycle?" http://soilcarboncenter.k-state.edu/carbcycle.html

Stokstad, E. 2011. "Can biotech and organic farmers get along?" *Science* 332(6026):166–169.

Sundquist, E. T., K. V. Ackerman, L. Parker, and D. Huntzinger. 2009. "An introduction to global carbon cycle management." In *Carbon Sequestration and Its Role in the Global Carbon Cycle,* ed. B. J. McPherson and E. T. Sundquist. Washington, DC: American Geophysical Union.

Van Oost, K., H. Van Hemelryck, and J. W. Harden. 2009. "Erosion of soil organic carbon: implications for carbon sequestration." In *Carbon Sequestration and Its Role
in the Global Carbon Cycle*, ed. B. J. McPherson and E. T. Sundquist. Washington,
DC: American Geophysical Union.

Williams, J. H., A. DeBenedictis, R. Ghanadan, A. Mahone, J. Moore, W. R.
Morrow, S. Price, and M. S. Torn. 2012. "The technology path to deep greenhouse gas emissions cuts by 2050: the pivotal role of electricity." *Science*
335(6064):53–59.

Witze, A. 2011. "Environment: human climate meddling got started long before dawn
of petroleum era: clearing forests released greenhouse gases by the gigaton."
Science News 179(9):17.

World Bank. 2011. "Cookstoves." http://www.climatechange.worldbank.org/
content/cookstoves-report

World Stove. 2010. http://www.worldstove.com/about_2/pilot-programs/

3

Terrestrial Carbon, Food Security, and Biosequestration Enhancement

3.1 Land and Carbon Management

Land played a pivotal role in the emergence of the global industrial market economy and it remains essential to its transformation. Our relationship with and treatment of land is so fundamental to survival that it seems to have become invisible, like the scenery we pass each day can become so routine that we fail to take note unless something changes. Karl Polanyi, in *The Great Transformation*, recognized the importance of land to human existence and considered its conversion into a commodity as a defining feature of market economies:

> What we call land is an element of nature inextricably interwoven with man's institutions. To isolate it and form a market out of it was perhaps the weirdest of all undertakings of our ancestors.... Traditionally, land and labor are not separated; labor forms part of life, land remains part of nature, life and nature form an articulate whole. Land is thus tied up with the organizations of kinship, neighborhood, craft, and creed—with tribe and temple, village, guild and church.... The economic function is but one of many vital functions of land. It invests man's life with stability; it is the site of his habitation; it is a condition of his physical safety; it is the landscape and the seasons. We might as well imagine his being born without hands and feet as carrying on his life without land. And yet to separate land from man and organize society in such a way as to satisfy the requirements of a real estate market was a vital part of the utopian concept of a market economy.*

We have become so accustomed to land as real estate, as a commodity, that it is easy to forget that we are utterly dependent on land and its covering of soil for our survival. Polanyi's characterization of a market economy as a utopian concept provides a hint of his analysis of free markets as a hopelessly temporary basis upon which to organize an economy (the

---

* Karl Polanyi, *The Great Transformation* (Boston: Beacon Press, 1957), 178.

concept of unsustainability was still unheard of in Polanyi's time). The industrial market economy of the nineteenth and twentieth centuries produced an unprecedented material abundance for those who appropriated the land of others. The appropriation of land in areas colonized by England and other European powers, the agricultural capitalism of the Tudors in England itself, and the land grabs of industrial capitalism for "industrial towns with their need for practically unlimited food and raw material supplies" were all "stages in the subordination of the surface of the planet to the needs of industrial society."[*] The appropriation of land continues and has accelerated recently, with China, India, other rapidly industrializing nations, and private investors buying land, primarily in Africa and Latin America, to meet the demands of economies that have exceeded the ecological resources of their homeland. This modern extension of enclosure and resource expropriation is driven largely by land and water shortages and the need to increase food production. [†]

The subordination of the surface of the planet, as noted by Polanyi, has led to the nearly universal degradation of the terrestrial ecosystems of the Earth. Reversing that degradation and restoring terrestrial ecosystems requires an approach to land use and management that eschews subordination and embraces careful recognition of the fragility, resilience, ecological limits, and unique characteristics of each biome and location—a distinctly nonindustrial approach. Rather than the simplification, mechanization, concentration, consolidation, and energy intensification associated with industrial land management, ecological land management involves approaches that recognize the complexity and uniqueness of each place, avoid concentration of ownership and control, and transition away from fossil fuels and toward the use of diffuse, renewable energy sources. Replenishing the organic matter and nutrients in soils that have been nearly universally diminished by agricultural and other land use and management practices is and will be a central component of improved practices. The large amounts of carbon that have been lost from forest, crop, and grazing lands worldwide present a historically unique opportunity to sequester large quantities of carbon over the next few decades by replacing the carbon lost due to a 10,000-year history of carbon-depleting practices.[‡]

---

[*] Polanyi, *The Great Transformation*, 179.

[†] Joachim von Braun and Ruth Suseela Meinzen-Dick, "'Land grabbing' by foreign investors in developing countries: risks and opportunities" (IFPRI Policy Brief 13, Washington, DC, International Food Policy Institute, 2009).

[‡] Rattan Lal, "Carbon sequestration," *Philosophical Transactions of the Royal Society B: Biological Sciences* 363, no. 1492 (2008); Rattan Lal, "Sequestering carbon in soils of agro-ecosystems," *Food Policy* 36, Supplement 1 (2011); Yude Pan et al., "A large and persistent carbon sink in the world's forests," Science 333, no. 6045 (2011); Wes Jackson, *Consulting the Genius of the Place: An Ecological Approach to a New Agriculture* (Berkeley, CA: Counterpoint Press, 2010).

There are a couple clear and universal principles underlying land use and land management practices that enhance the biosequestration of carbon; all involve increasing and protecting photosynthetic biomass or increasing the amount of carbon stored in soil. Increasing the amount of carbon stored in soil results from increased carbon inputs from aboveground and below-ground biomass, reduced carbon losses due to disturbance and exposure, or a combination of these. In simple terms, this means growing more and healthier plants and keeping soil covered with organic matter. The carbon in biomass and soils is most stable and secure when produced and protected in a healthy ecosystem with a diversity of perennial species and a minimum of disturbance. Carbon and nutrients in cropland that is disturbed by tillage can be protected and enhanced by minimizing fallow periods, using cover crops, and using multicrop systems.

Biological sequestration of carbon in soil has multiple benefits in addition to removing it from the overloaded atmosphere—it enhances fertility, increases water retention capacity, and improves primary production. Increasing the biomass and soil carbon on a parcel of land with a diverse range of perennial plant species that are native or well adapted to the location enhances eco-logical diversity and resilience to extreme weather events such as drought and flooding. Enhancing biosequestration of carbon makes an essential con-tribution to both mitigation and adaptation to global warming and climate change. New and emerging evidence indicates that this contribution has the potential to be even larger and more significant than recognized even a few years ago.[*]

Biosequestration of carbon is accomplished most effectively, both in terms of quantity of carbon and its relative permanence, in healthy, diverse ecosystems. By reducing the atmospheric concentration of car-bon dioxide, terrestrial biosequestration also reduces the extent of ocean acidification, sea level rise, and many of the associated ongoing degrada-tions of marine ecosystems. Human land use during the Anthropocene, especially agriculture and the deforestation resulting from its expan-sion, has been and remains the primary driver of global biodiversity loss. It appears that we are about to enter a new phase of human impact, as anthropogenic climate change may equal or exceed land use as a driver of biodiversity loss in coming decades.[†] Fundamentally changing the way we use and manage forests, grasslands, cropland, wetlands, and other ter-restrial ecosystems can slow the rate, degree, and impact of biodiversity loss and climate change. Actions to enhance biosequestration and build soil meet the test of addressing more than one problem; they directly con-tribute to mitigating many of the central ecological degradation problems of the Anthropocene.

---

[*] Lal, "Sequestering carbon in soils of agro-ecosystems."
[†] Terence P. Dawson et al., "Beyond predictions: biodiversity conservation in a changing cli-mate," *Science* 332, no. 6025 (2011).

3.2 Forest Biomes and Carbon Sinks

The largest carbon sequestration and ecological restoration potentials exist in the world's forests. There are 148.94 million square kilometers (57.49 million square miles) of land area on Earth. In 2010, forests covered about 40.5 million square kilometers (15.63 million square miles) or approximately 27% of the Earth's land area.[*] Oceans cover 70.8% of the Earth's surface, so forests constitute about 8% of the total surface area of the planet. This small fraction of the Earth's surface plays a key role in the global carbon cycle as the largest of the terrestrial carbon sinks. It also provides a vast range of essential ecological services to human societies and millions of plant and animal species that live in or depend on forests for their survival.

As noted earlier, there has been substantial uncertainty and a range of estimates regarding the current capacity of the Earth's terrestrial ecosystems to sequester carbon. The Intergovernmental Panel on Climate Change (IPCC) has estimated net terrestrial carbon uptake at anywhere from less than 1 gigaton to as much as 2.6 gigatons per year.[†] More recent global analyses have estimated the net terrestrial carbon sink to be in the range of 2–3.4 gigatons per year.[‡] In 2011, Pan et al. published a comprehensive new analysis in *Science* of the living biomass, deadwood, harvested wood products, litter, and soil (to a depth of 1 meter) in the Earth's forests.[§] This research provides important new evidence about the largest of the terrestrial system's carbon sinks. Pan confirmed that, despite some remaining uncertainties, the carbon sequestration taking place in the Earth's forest ecosystems has increased since 1990. The data compiled by Pan et al. document the carbon sequestration contribution made by intact boreal, temperate, and tropical forests as well as that made by the regrowth of tropical forests recovering from human disturbance.

There was a persistent global gross carbon sink of 4.05 (±0.67) gigatons per year from 1990 through 2007 in the Earth's forests. This includes all of the carbon sequestered in biomass and soil by intact boreal, temperate, and tropical forests as well by regrowth of tropical forests following anthropogenic disturbances. Deforestation of intact tropical forests, however, has been a source of about 2.94 (±0.47) gigatons of carbon emissions per year for the same 1990–2007 period. This means that the global net forest sink has averaged 1.11 (±0.82) gigatons of carbon sequestration per year during the past two decades. This inventory shows that although deforestation of the world's tropical forests is continuing, it has slowed in recent years. During the 1990–1999 period the net tropical land use emission averaged 1.46 (±0.7)

---

[*] Pan et al., "A large and persistent carbon sink in the world's forests."
[†] IPCC, *Climate Change 2007: Synthesis Report*, Fourth Assessment Report (AR4) (Geneva: Intergovernmental Panel on Climate Change, 2007).
[‡] Pan et al., "A large and persistent carbon sink in the world's forests."
[§] Pan et al., "A large and persistent carbon sink in the world's forests."

gigatons per year and declined to 1.10 (±0.7) gigatons per year in the 2000–2007 time period. If accurate, this is a 25% decline in tropical deforestation carbon emissions over a short time frame. A close look at the data reveals that the measured decrease in net emissions from tropical forests is due to both a higher rate of carbon sequestration in tropical regrowth forests and a declining rate of deforestation emissions from intact tropical forests.[*]

This new evidence has profoundly positive implications for our potential ability to address the climate and carbon imbalance. By continuing to slow the rate of deforestation in tropical forests we can potentially negate a significant and increasing portion of global carbon emissions from fossil fuels. If gross emissions from tropical deforestation could be reduced by one-third from the annual average during the 2000–2007 period, the net annual carbon sequestration of the world's forests would increase by about 0.94 gigatons. Pan et al. estimate that net carbon accumulation in the Earth's atmosphere was 4.1 gigatons per year during the 2000–2007 period; a 0.94 gigaton decrease in tropical forest emissions, all other factors being equal, would reduce the annual accumulation of carbon in the Earth's atmosphere by nearly one-fourth. By eliminating tropical deforestation entirely and maintaining current levels of sequestration in boreal, temperate, and tropical regrowth forests, the gross forest sink of more than 4 gigatons of carbon per year could possibly become the net carbon sink of the Earth's forests. This level of sequestration in forests could reduce the net accumulation of carbon in the atmosphere from the recent 4.1 gigatons per year to about 1.28 gigatons per year, a reduction of about 69%. Net carbon sequestration of the Earth's forests could increase further if the increasing rate of carbon sequestration in tropical regrowth forests and temperate forests continues. This is likely an overly optimistic scenario, but it demonstrates that the global magnitude of the carbon in the Earth's forests is indeed significant in relation to the net annual carbon accumulations in the atmosphere.

Total elimination of tropical deforestation will not occur overnight. Making progress toward this goal will require coordinated international cooperation and investment and changing deeply entrenched agricultural, forestry, mining, and settlement practices. Beyond direct tropical deforestation, persistent warming in boreal (northern) areas and increasingly frequent droughts and associated fires in the tropics represent the greatest threat to preservation of the large carbon sink in the world's forests.[†] If current trends toward more extreme droughts, fires, and insect infestations in temperate and boreal zones continue, it will also be difficult to maintain the currently increasing rate of carbon sequestration in temperate forests and the stable carbon stock in boreal forests.

---

[*] Pan et al., "A large and persistent carbon sink in the world's forests."
[†] Ibid.

Land use and management actions to reduce the rate of tropical deforestation and to carefully manage and expand boreal, temperate, and tropical regrowth forests have probably the largest potential of any set of actions to enhance and protect the global terrestrial carbon sink. To better understand the current carbon stock in global forests and the areas of greatest risk and opportunity for improvement, it is helpful to look at the carbon stocks, land areas, and trends in each of the forest types (biomes).

The total area of forested land was reduced by about 3.4% from 1990 to 2010.[*] This is a decline of about 1.35 million square kilometers (522,400 square miles), an area larger than the Canadian province of Ontario and the state of Minnesota combined. The rate of global deforestation averaged more than 82,000 square kilometers (32,000 square miles) per year during the 1990s and declined to about 51,000 square kilometers (20,000 square miles) per year between 2000 and 2010.[†] The global decline in forested area was entirely from the loss of tropical forest, which exceeded overall gains in both boreal and temperate forest area.[‡]

3.2.1 Boreal Forests

Boreal (northern) forests cover about 1,300 million hectares (3,212 million acres) or about 5 million square miles of land area and store about 272 gigatons of carbon. Boreal forests make up about 32% of the Earth's forest area and store approximately 32% of the carbon in global forests. Boreal forests store 60% of their carbon in soil and only 20% in live biomass, with the remainder in harvested wood products, deadwood, and litter. In contrast, tropical forests have a fundamentally different carbon structure, with 56% of their carbon in live biomass and 32% in soil.[§] This difference means the carbon in tropical forests, predominately stored in live biomass, is especially vulnerable to disturbance. Boreal forests store more carbon in the soil, where it is relatively more secure.

Between 1990 and 2007 boreal forests added to their carbon stock by sequestering about 0.5 gigatons of carbon per year. Unlike both temperate and tropical forests, boreal forests have not appreciably increased or decreased their net rate of carbon sequestration over the past two decades despite a small increase in total land area. This global stability masks substantial changes occurring in some areas. Asian Russia has shown no net changes in its boreal carbon sink since 2000, while the sink increased by 35% in European Russia and other parts of northern Europe. The increasing rate of sequestration in European boreal forests is attributed to increased forest area due to agricultural abandonment, reduced harvesting, and a

---

[*] Food and Agriculture Organisation, *State of the World's Forests* (Rome: United Nations, 2011).
[†] Ibid.
[‡] Pan et al., "A large and persistent carbon sink in the world's forests."
[§] Ibid.

change in forest age structure to a more productive stage. The increased rate of sequestration in European boreal forests was partially offset by a large 50% reduction of carbon uptake in Canada's managed boreal forests. The Canadian boreal sequestration decline was due primarily to biomass loss from intensified wildfires and insect damage, both related to warming. The net increase in carbon in the form of biomass in northern Europe would have been larger, but was partially offset by a net loss of soil carbon there due to draining of wetlands.*

There are several potential opportunities for enhancement and protection of boreal carbon stocks. The management of boreal forests and their climate-related impacts in two nations, Russia and Canada, will largely determine the future of the global boreal forest biome. Russia has by far the largest forested land area on the Earth, with more than 809 million hectares (1,999 million acres), nearly all of it boreal. Canada has the third largest forested land area of all nations, at more than 310 million hectares (766 million acres).† Canada's forest is also almost exclusively boreal. The carbon sink in Asian Russia is the largest of the boreal areas. Although it was stable overall during the research period examined by Pan et al., this area, like Canadian managed forests, experienced increased fire disturbance,‡ which reduced the rate of regional carbon uptake. Additional research and resources devoted to wildfire prevention and management in boreal forests could reduce the damage and carbon emissions due to wildfires. As global warming continues to warm northern latitudes at rates up to twice global averages, the risks from both wildfires and insect outbreaks will likely intensify. This increased level of risk presents a complex management challenge.

Soil drying and carbon losses can be minimized in areas disturbed by wildfires, insect damage, and harvesting by replanting as rapidly as possible. This is especially critical in boreal forests, where a high percentage of total carbon is stored in soils and live biomass is relatively slow growing. Rapidly reestablishing cover to disturbed areas protects carbon in soils from drying and reduces rates of carbon emissions. Controlled burning can also play a positive role in enabling conifers to reproduce, especially the long-lived and slow-growing Jack pine. Loss of biomass and the threat posed to exposed soil carbon from uncontrolled wildfire has reduced the carbon sink of both the Canadian and Russian boreal forests and will likely continue to increasingly threaten the longevity of sequestered carbon in these areas.

The loss of soil carbon in the boreal forests of northern Europe due to the draining of wet soils can be more effectively avoided by human management than the damage caused in other areas by fire and insect infestations related to climate change. This, however, requires recognition

---

* Pan et al., "A large and persistent carbon sink in the world's forests."
† Food and Agriculture Organisation, *State of the World's Forests.*
‡ Pan et al., "A large and persistent carbon sink in the world's forests."

that the ecological value of the carbon lost by draining wet boreal soils may exceed the market value of the land use made possible by draining. Conventional economic thinking that treats ecological damage as an externality has resulted in a common failure to place sufficient value on the primary products, like soil carbon, of nature. This is changing among some economists, policymakers, and others who have recognized that the ecological services of nature are the foundation underlying all secondary products created by humans. This recognition and its incorporation into land use and land management policies and practices can slow and reverse the draining of boreal wetlands and a range of similarly short-sighted actions. This is a concrete example of the kinds of changes involved in evidence-based improvements to land management practices. Underlying such changes is a shift away from a paradigm that pursues economic growth without recognition or adequate consideration of ecological limits, costs, and consequences.[*]

About 1.9 million acres of trees are cut per year in the Canadian boreal forest and approximately 65% of the trees are used for pulp and paper. In North America, much of that virgin pulp is used to print some 20 billion catalogs per year. Those catalogs could instead be made from recycled office paper and recycled newspapers. Kimberly-Clark, the maker of Kleenex, uses more than 500,000 tons per year of virgin pulp to make toilet paper, facial tissues, napkins, and paper towels. Other manufacturers have demonstrated that these products can be made with recycled paper without compromising quality.[†] Reducing the use of disposable paper products and making them from recycled paper helps keep the carbon in boreal forest trees and soil and out of the atmosphere.

Paper recycling must be done in an ecologically sound manner, however, to have the greatest benefits. This has implications for the entire paper products industry. Chlorine and other chemicals are often used in the production of recycled paper products and there are also toxic chemicals in many inks used in printing. Ecologically sound paper recycling requires eliminating the use of dangerous chemicals in the recycling process and the elimination of toxic chemicals used in printing, whether the paper is made from virgin pulp or recycled material.[‡] Long-lasting harvested wood products used for construction and furniture can sequester carbon for decades to centuries. Finding ways to substitute renewable wood for fossil fuel–intensive materials such as plastic, concrete, and steel can both reduce fossil fuel use and enhance long-term carbon sequestration.

---

[*] John Michael Greer, *The Wealth of Nature: Economics as if Survival Mattered* (Gabriola, BC, Canada: New Society Publishers, 2011).

[†] T. Edward Nickens, "Paper chase," *Audubon* January–February (2009).

[‡] William McDonough and Michael Braungart, *Cradle to Cradle: Remaking the Way We Make Things*, 1st ed. (New York: North Point Press, 2002).

3.2.2 Temperate Forests

Temperate forests cover about 800 million hectares (1,977 million acres) or about 3.1 million square miles and store approximately 119 gigatons of carbon. Temperate forests constitute about 20% of global forest area and hold about 13% of the carbon in forests. The carbon density of temperate forests is substantially lower than boreal and tropical forests, with about 155 metric tons of carbon per hectare (2.47 acres) compared to 242 and 239 metric tons per hectare for tropical and boreal forests, respectively. Temperate forests have contributed an increasing amount of carbon sequestration over the past twenty years because of an increasing density of biomass and a substantial increase in forest area. Temperate forests added to their global sink by sequestering about 0.7 gigatons of net carbon per year from 1990 through 1999 and increased to about 0.8 gigatons per year from 2000 to 2007.[*]

The size of the carbon sink in temperate forests in the United States increased by about one-third from 1990 to 2007. The increase was caused by increasing forest area, especially in the eastern states, where immature forests are growing rapidly as they recover from agricultural abandonment, extensive grazing, and harvesting. Many forested areas in the western United States have been under severe stress from extensive droughts, wildfires, insect infestations, and a general increase in mortality.[†] Much of the western United States, especially the southwestern portion, may be entering a long-term period of aridification or drying. The frequent droughts experienced in recent years in the southwestern United States may be the new normal for that region.[‡] Lower rainfall in the southwestern United States and northern Mexico over the long term will likely contribute to a continued deterioration of forest stand quality, with negative implications for overall carbon sequestration and emissions. Despite problems with the health of western forests, the rapid growth in the eastern United States has so far produced net increases in carbon sequestration.

The carbon sink in the temperate forests of China also expanded by about one-third from 1990 through 2007, with the carbon in biomass nearly doubling. Most of the increase in temperate forest carbon in China during the past two decades is from a large increase in forested area due to intensive afforestation and reforestation programs.[§] The large increase in temperate forest area and temperate forest carbon storage in both the United States and China is significant globally, as these two nations have the largest extent of

---

[*] Pan et al., "A large and persistent carbon sink in the world's forests."
[†] Phillip J. van Mantgem et al., "Widespread increase of tree mortality rates in the western United States," *Science* 323, no. 5913 (2009).
[‡] Richard Seager et al., "Model projections of an imminent transition to a more arid climate in southwestern North America," *Science* 316, no. 5828 (2007).
[§] Pan et al., "A large and persistent carbon sink in the world's forests."

temperate forests in the world. Together they contain more than 60% of the Earth's temperate forests.[*]

As noted previously, the temperate forests of the southwestern United States are under threat from a persistent aridification due to climate change. The large area of temperate forests in northern Mexico is affected by the same change in precipitation patterns. In western Asia the largest temperate forests are found in Turkey, where there has been a small but persistent increase in both forested area and carbon storage since 1990. South of the Tropic of Capricorn, Chile has experienced a small but persistent increase in forested area, while southern Australia has had a small recent decline.[†] The net affect has been a global increase in temperate forest extent and carbon due primarily to the large expansions of forested area and carbon density in Europe, the United States, and especially China.

It remains to be seen whether this growing contribution to the global carbon budget can be sustained. It appears likely that China will hold the key to whether global temperate forests continue to expand in coming decades. The total Chinese forested area expanded at a rate of nearly 3 million hectares (7.4 million acres) per year from 2000 through 2010. To put this in perspective, China's annual increase in forested area exceeds the total annual area of forest expansion in all of the other nations of the world that managed to expand their forest area since 2000. The United States had the next largest annual increase, at 383,000 hectares (946,400 acres). India was third, with 304,000 hectares (751,200 acres), and Viet Nam was fourth, with 207,000 hectares (511,500 acres) per year.[‡] The added forest areas in India and Viet Nam are mostly tropical forest, with a higher carbon density and greater ecological diversity than the mostly temperate forest plantations in China. The sheer size of the Chinese temperate forest area and their afforestation and reforestation projects, however, make it likely that China will remain the single most important nation in affecting the global temperate forest carbon sink in the immediate future.

3.2.3 Tropical Forests

Tropical forests are the largest global forest biome in both area and carbon storage. They cover about 1,950 million hectares (7.53 million square miles) and store about 471 gigatons of carbon. Tropical forests account for about 48% of global forest area and approximately 55% of the carbon in forests. Although tropical forests store slightly more carbon per hectare than boreal forests (242 compared to 239 metric tons per hectare, respectively) and much more than temperate forests (155 metric tons per hectare) they are a net

---

[*] Food and Agriculture Organisation, *State of the World's Forests.*
[†] Ibid.
[‡] Ibid.

carbon emitter overall. This is because tropical deforestation is second only to fossil fuels as a source of carbon emissions to the Earth's atmosphere.

From 2000 through 2007 fossil fuel and cement emissions were, according to Pan, about 7.6 gigatons per year and gross emissions from tropical deforestation were about 2.82 gigatons per year. The rate of gross emissions from tropical forests has declined from an average of 3.03 gigatons per year for the 1990–1999 period, while the emissions from fossil fuels and cement have increased to 7.6 gigatons per year from an annual average of 6.5 gigatons during the 1990s.[*] The 7.6 gigaton estimate of annual fossil emissions provided by Pan in 2011 may be conservative. The American Geophysical Union published an estimate of 8.5 gigatons of carbon emissions per year as of 2007,[†] and the approximately 30 gigatons of carbon dioxide emissions reported by the International Energy Agency for 2010 translates to 8.19 gigatons of carbon. These figures provide a reasonable estimate of relative magnitude.

Despite the large potential of global tropical forests for annual carbon sequestration and the large amount of carbon that is currently stored in them, the net annual carbon sink (addition) of 1.2 gigatons in global forest ecosystems in recent years is due to the sustained net sequestration of the Earth's boreal and temperate forests. As noted, forests have the greatest potential among all terrestrial biomes for enhanced biosequestration of carbon. Among forest biomes, tropical forests have the greatest potential for enhanced net sequestration. Realizing more of this potential requires continuing to slow and eventually eliminating tropical deforestation.

The largest tropical deforestation since 1990 has occurred in Brazil, the nation with by far the largest extent of tropical forest. Brazil, with more than 519.5 million hectares (1,283.7 million acres) of forest land, is second only to Russia in total forested area, and nearly all of it is in the tropics. Brazil lost an average of about 2.9 million hectares (7.2 million acres) of forest area per year from 1990 through 2000. The rate of deforestation in Brazil declined by about 10%, to 2.6 million hectares (6.4 million acres) per year, from 2000 through 2010. In all of South America the rate of deforestation has also slowed slightly, from 4.2 million hectares (10.4 million acres) per year in the earlier period to 4.0 million hectares (9.9 million acres) per year since 2000.[‡]

After South America, the largest area of tropical forest is in Africa. Tropical deforestation in Africa is dispersed among many nations across the continent. Some of the larger losses have been in Tanzania and Ethiopia in east Africa; the Congo and Cameroon in central Africa; Sudan (which has drastically reduced its rate of deforestation in recent years) in northern Africa; Zimbabwe, Mozambique, Zambia, Angola, and Botswana in southern Africa; and Nigeria and Ghana in west Africa. The rate of deforestation on

---

[*] Pan et al., "A large and persistent carbon sink in the world's forests."

[†] Eric T. Sundquist et al., "An introduction to global carbon cycle management," in *Carbon Sequestration and Its Role in the Global Carbon Cycle*, ed. Brian J. McPherson and Eric T. Sundquist (Washington, DC: American Geophysical Union, 2009).

[‡] Food and Agriculture Organisation, *State of the World's Forests*.

the African continent has also declined, from an average of 4.1 million hectares (10.1 million acres) per year during the 1990s to 3.4 million hectares (8.4 million acres) per year since 2000.[*]

In Asia the majority of tropical deforestation has occurred in the Southeast Asian nations of Indonesia, Myanmar, Cambodia, and Malaysia. Although Indonesia has continued to experience the largest deforestation in Asia since 2000, the rate of annual deforestation there has been reduced by nearly 75% since the 1990s. This and a reduced rate of deforestation in Myanmar have been the primary reasons that the overall rate of tropical deforestation in Southeast Asia has been reduced by about 63%, from an average of 2.4 million hectares (5.9 million acres) per year during the 1990s to 0.9 million hectares (2.2 million acres) per year since 2000. In south Asia, India has a mix of tropical forest in the south and temperate forests in the north and has added forest area at a steadily increasing rate since 1990. Since 2000 India has added an average of 304,000 hectares (751,000 acres) of forest area per year. One final large land mass, Australia, also has a mix of tropical and temperate forests. Australia experienced a large reversal from a modest net gain of 42,000 hectares (104,000 acres) in forest extent per year during the 1990s to a large net deforestation of 562,000 hectares (1.4 million acres) per year since 2000.[†] Australia appears to be in a long-term aridification process due to climate change similar to that occurring in the southwestern United States.[‡]

3.2.4 A Precarious Improvement

Boreal forests have remained relatively stable in extent (area) since 1990 and have been steadily sequestering about 0.5 gigatons of net carbon per year through 2007, the latest year for which data are available. Temperate forests have increased in extent due primarily to the large afforestation/reforestation program in China and to a lesser extent due to widespread reforestation in Europe, the United States, and India. Temperate forests have increased their rate of net carbon sequestration to about 0.8 gigatons per year since 2000 from about 0.7 gigatons during the 1990s. Although tropical forests are the largest and most carbon dense of the global forest biomes, deforestation and forest degradation primarily in South America, Africa, and Southeast Asia continue to make tropical forests a net emitter of carbon.

The bottom line on forest carbon is that since 2000 the Earth's forests have provided a gross sink of about 4 gigatons of carbon per year and a net sink of about 1.2 gigatons per year when gross tropical deforestation emissions of about 2.82 gigatons are subtracted. This is a slightly larger net carbon sink than that provided by global forests during the 1990s because the rate of tropical deforestation

---

[*] Food and Agriculture Organisation, *State of the World's Forests*.
[†] Ibid.
[‡] Seager et al., "Model projections of an imminent transition."

has declined, tropical forest regrowth has increased, and the extent of temperate forests has expanded. These global trends are moving in a positive direction for now. The inherent instability of terrestrial carbon, especially in living biomass, coupled with climate change–driven increases in drought, wildfires, and insect infestations, make the positive trend in forest carbon precarious at best. Widespread drought in the Amazon rain forest in 2005 and 2010, for example, resulted in carbon emissions of 1.6 gigatons and 2.2 gigatons, respectively, undoing much of the progress made over the past ten years.[*]

Careful monitoring and management at local, regional, and global levels are improving and will be required on an ongoing basis to maximize and preserve forest carbon sinks. Nearly everyone can participate by planting and nurturing trees unless doing so will interfere with food production or cause ecological or infrastructure problems. This admonition applies to urban, suburban, and rural areas anywhere from about 40 degrees south (central Argentina) latitude to at least 60 degrees north latitude (which runs through the Hudson Bay).

3.2.5 Tropical Deforestation and Fossil Fuel Emissions

If the general magnitude of the global forest carbon sink documented by Pan et al. and their estimate of net annual carbon accumulation in the atmosphere are correct and confirmed by further research, the recent net global forest carbon sink of 1.2 gigatons is reducing the annual accumulation of carbon in the Earth's atmosphere by nearly one-third compared to what it would be without the carbon services of the Earth's forests. The 4.1 gigatons of net annual carbon accumulation in the Earth's atmosphere estimated by Pan et al. would be about 5.3 gigatons without the net contribution made by global forest ecosystems. By either increasing the gross forest sink, reducing the emissions due to tropical deforestation, or a combination of both, the net contribution made by the Earth's forests can grow. To gain a sense of the total potential magnitude of forest carbon sequestration, consider an admittedly optimistic scenario: eliminating all 2.82 gigatons of annual carbon emissions due to current rates of tropical deforestation would reduce the current annual net accumulation of carbon in the atmosphere by about two-thirds, even if fossil fuel emissions stay at current levels. This is not likely to happen quickly because the complex social forces driving deforestation are, like fossil fuels, deeply enmeshed in the central features of industrialism: the growth imperative and a failure to recognize and account for the full value of the inherently finite products of nature and functioning, healthy ecosystems.[†] Every net gigaton of increased carbon sequestered by the Earth's forests, however, reduces by 1 gigaton the net accumulation of carbon in the atmosphere.

---

[*] Nina Chestney, "Global warming close to becoming irreversible-scientists," *Breaking US and International News*, http:/www.reuters.com/assets/print?aid=USBRE82POUJ20120326

[†] Greer, *The Wealth of Nature*; Richard Heinberg, *The End of Growth: Adapting to Our New Economic Reality* (Gabriola, BC, Canada: New Society Publishers, 2011).

The causes of tropical deforestation, as expected in the Anthropocene, are human activities. These include agricultural expansion of cattle grazing and cropland; harvesting trees for timber and fuel; mining and oil extraction; construction of dams, roads, settlements, and other infrastructure; and climate change.[*] The growing demand for beef, wood, and mineral ores must be met or, preferably, reduced in ways that do not require ongoing deforestation. The policies, politics, and economic transformations and land use changes needed to make that possible are emerging, mostly in the global south.[†]

Meantime, it is sobering to recognize that despite measurable improvements in global forest carbon sinks over the past two decades, these improvements have been overwhelmed by ongoing increases in emissions from fossil fuels. Fossil fuel emissions increased from at least 6.5 gigatons of carbon per year during the 1990s to at least 7.6 gigatons per year since 2000. The most important bottom line, in terms of carbon, is that the net accumulation in the Earth's atmosphere is still increasing. Pan et al. estimated that the net uptake of carbon by the atmosphere was 4.1 gigatons per year from 2000 through 2007, up from 3.2 gigatons per year during the 1990s.[‡] The situation has not improved since 2007, as global emissions declined slightly during the global economic slowdown of 2008–2009 and returned to record levels in 2010.

The positive news is that annual net carbon accumulation in the Earth's atmosphere is clearly less than it would be if not for the persistent and recently growing carbon sink in global forests. This represents an important contribution to mitigating global warming and climate change. As we have seen, one of the greatest uncertainties regarding the future rate and magnitude of warming and climate change revolves around the degree to which terrestrial ecosystems continue to act as a net carbon sink. Actions taken to enhance and preserve the health of forest ecosystems, by implementing proven forest management practices, can help preserve the largest of the terrestrial carbon sinks and, at the same time, provide a wide range of related ecological benefits beyond carbon.

Carbon, of course, is critically important and will be generally maximized over the long term by appropriate forest management (which can mean leaving some natural processes "unmanaged"). But carbon is not the only ecological benefit of forests, and preserving forest carbon sinks for long-term stability requires more than just trees. Healthy, productive, and long-lasting forest ecosystems both require and provide a habitat with a rich diversity of plant and animal communities. Intensive plantation management to produce the greatest volume of wood products in the shortest possible amount of time at the highest profit margin can, and almost certainly will, fail to provide the species diversity and soil organic carbon from litter and decayed

---

[*] Helmut J. Geist and Eric F. Lambin, "Proximate causes and underlying driving forces of tropical deforestation," *BioScience* 52, no. 2 (2002).

[†] Kolya Abramsky, *Sparking a Worldwide Energy Revolution: Social Struggles in the Transition to a Post-Petrol World* (Oakland, CA: AK Press, 2009).

[‡] Pan et al., "A large and persistent carbon sink in the world's forests."

deadwood required to maintain healthy forest ecosystems.[*] An ecological perspective on forests must recognize that sufficient quantities of decaying woody debris, a diversity of species, a full range in the ages of trees, a great diversity of other plant, fungi, and animal species, and a healthy community of soil organisms are required to maintain and build healthy, fertile soils and forests. A plantation of one or two species of trees, all the same age, with woody debris largely removed, cannot support a sufficient diversity of plant and animal species or provide the organic matter required for healthy soils and a stable long-lasting forest ecosystem. Yet this is a fair description of many of the managed forests throughout the world.[†]

3.2.6 Ecological Restoration

Protecting the intact natural forests that still exist in every biome will preserve both long-lasting carbon sinks and some of the most biologically diverse ecosystems on Earth. Afforestation, reforestation, and ecological restoration projects can be, and increasingly are, designed and managed to re-create as much diversity as possible. The core principle of ecological forest restoration is to "enhance ecological integrity by restoring natural processes and resiliency."[‡] This process requires careful planning and a comprehensive, multidisciplinary scientific assessment of existing and historic conditions at regional and site-specific scales. Ecological restoration, whether in forests or any other terrestrial ecosystem, involves a combination of protection and active or passive restoration strategies. Identifying and protecting relatively intact natural areas that have a high degree of ecological integrity and need little restoration is essential as a reference or baseline by which to gauge the restoration of more degraded landscapes and as a valuable source of biodiversity. The primary objective of restorative practices is to maintain or reestablish fully functioning ecosystems that support a resilient community of diverse organisms. After protection of relatively intact ecosystems, the least invasive and lowest-cost restoration strategy is passive restoration, which involves stopping the activities that cause degradation. This allows natural processes of recovery to occur and often involves halting destructive logging, grazing, mining, road and other infrastructure building, and destructive fire regimes or excessive fire suppression.[§]

The principles that apply to forests also apply to sequestering carbon and building soil in native prairies, managed grasslands, croplands, and wetlands. A healthy soil requires a steady supply of plant tissue (carbon) to retain the capacity to support the primary production of photosynthesis. A very brief

---

[*] Chris Maser, *Sustainable Forestry: Philosophy, Science, and Economics* (Boca Raton, FL: St. Lucie Press, 1994).

[†] Maser, *Sustainable Forestry*.

[‡] Dominick A. DellaSala et al., "A citizen's call for ecological forest restoration: forest restoration principles and criteria," *Ecological Restoration* 21, no. 1 (2003): 15.

[§] DellaSala et al., "A citizen's call for ecological forest restoration."

foray into the complex interactions that take place below the soil surface is instructive. Photosynthesis above the soil and related multispecies and nutrient (diversity again) interactions below the terrestrial surface make biosequestration of carbon an essential (and fascinating) link in the web of terrestrial life. One of the more essential below-the-surface processes involved in biosequestration requires an equitable exchange between plants and fungi. Carbon extracted from the atmosphere by plants, combined with hydrogen and oxygen to form carbohydrates, is the currency that forms the basis for complex symbiotic trades between nearly all terrestrial plants and arbuscular mycorrhizal fungi. Mycorrhizal fungi provide phosphorus and other mineral nutrients to plants in direct exchange for carbohydrates. These carbohydrates are as essential for the growth of fungi as mineral nutrients are to plants. "This partnership, which evolved long before mutualisms among insects or vertebrates, is credited with driving the colonization of land by plants, enabling massive global nutrient transfers and critical carbon sequestration."[*]

A large part of our task is to allow such natural ecological processes to flourish by nurturing carbon- and species-rich ecosystems unpolluted by synthetic chemicals. Such ecosystems have become scarce, but fragments exist in boreal, temperate, and tropical forest biomes. Fragmentation of forest ecosystems, increasing drought, and more frequent pest and disease infestations have had a disproportionate impact on the largest trees in forest ecosystems, which are dying rapidly in many parts of the world. Large trees in fragmented forests are more exposed to high winds than in larger tracts. Small fragments of forest ecosystems are also more susceptible to drought. The largest trees in a forest ecosystem serve as an important source of seed and sequester up to 25% of the forest's carbon in biomass.[†] Preservation of fragments of healthy diverse forest ecosystems is only a starting point for forest restoration on a scale that can protect the carbon and the ecological diversity found in the Earth's forests. Truly healthy forests are not highly fragmented.

3.3 Agricultural Land, Degraded Soils, and Water Scarcity

3.3.1 Restoring Carbon to Agricultural Soils

The agricultural soils of the world have been a source of carbon emissions since the emergence of agriculture about 10,000 years ago.[‡] I noted earlier

---

[*] E. Toby Kiers et al., "Reciprocal rewards stabilize cooperation in the mycorrhizal symbiosis," *Science* 333, no. 6044 (2011): 880.

[†] John Vidal, "World's giant trees are dying off rapidly, studies show," *The Guardian*, http://www.guardian.co.uk/environment/2012/jan/26/giant-trees-dying

[‡] R. Lal et al., "Soil carbon sequestration to mitigate climate change and advance food security," *Soil Science* 172, no. 12 (2007).

that recent ice core data has provided evidence of altered methane emissions detectable in the atmosphere as early as 8,000 years ago, associated with the large-scale irrigated agriculture practiced in much of Asia at that time. Agriculture has transformed the ecosystems of the planet more than any other human activity except perhaps the extraction and burning of fossil fuels. This is true not only because of the damage done directly to cropland and pastures, but also because the expansion of agricultural land has been the primary driver of deforestation.

Industrial agriculture and fossil fuels are now integrated as closely as electricity production and fossil fuels. Indeed, weaning modern agriculture from fossil fuels may be one of the more difficult but necessary tasks of the next few decades. Along with breathable air and potable water, what is more necessary than food? The vast majority of contemporary agricultural practices are highly dependent on fossil fuels. In industrial agriculture, tilling, planting, weed and pest control, harvesting, drying, transporting, feeding livestock, processing, packaging, preserving, marketing, storing, and cooking of agricultural products are nearly universally done with machines that are fossil fuel intensive in both their manufacture and operation. The ongoing transfer of industrial agricultural practices and technologies to nonindustrial ("developing") nations and peoples continues to spread this fossil fuel dependence to ever-larger percentages of the human population. This results in a global agricultural system that is a growing contributor to both of the primary sources of carbon emissions: fossil fuels and terrestrial ecosystem degradation. The rapid globalization of industrial capitalism since World War II has accelerated these trends to unprecedented levels.*

The ecosystem degradation and carbon emissions from soils due to agricultural activities preceded the burning of fossil fuels by thousands of years. Rattan Lal, of the Carbon Management and Sequestration Center at Ohio State University (Columbus, Ohio), has estimated that most agricultural soils have lost 25%–75% of the soil organic carbon they held prior to agricultural disturbance. This loss may be 10–50 metric tons per hectare (4–20.2 metric tons per acre), with greater losses on highly eroded and degraded soils. Lal and other soil scientists identify this long history of soil organic carbon losses as one of the greatest opportunities for enhanced carbon sequestration in the future. The low hanging fruit in the realm of soil organic carbon involves recarbonizing degraded soils. Lal estimates a global potential of from 1.2 gigatons to 3.1 gigatons of additional soil organic carbon sequestration per year by restoring degraded soils through adoption of restorative land uses and recommended management practices on crop, irrigated, range, and grasslands.†

---

* Robert J. Antonio, "Climate change, the resource crunch, and the global growth imperative," in *Current Perspectives in Social Theory*, vol. 26, ed. Harry F. Dahms (Bingley, UK: Emerald Group Publishing, 2009).
† Lal, "Sequestering carbon in soils of agro-ecosystems."

Restorative land uses noted by Lal include improving pasture with perennial species and nitrogen-fixing legumes and afforestation or reforestation of
highly degraded or steeply sloping land currently used for crops or grazing.
The recommended practices for improving carbon sequestration in cropland
include conservation tillage, mulch farming, cover crops, and integrated
nutrient management, including the use of manure, compost, and agroforestry. The rate of potential carbon sequestration ranges from 100 kilograms
per hectare per year to 1,000 kilograms per hectare per year (89 pounds per
acre to 892 pounds per acre) depending on soil types, climate, and specific
management practices.[*]

The primary products sought from agricultural land are, of course, food,
fodder, fiber, and, increasingly, fuel. Managing agricultural lands to restore
soil organic carbon to depleted soils increases agronomic productivity, especially on soils with extremely low carbon levels. Lal proposes that increasing the soil organic carbon pool in the root zone by 1 metric ton per hectare
(892 pounds per acre) in the agricultural soils of developing nations would
greatly increase the production of cereals, legumes, roots, and tubers and
have a positive impact on global food security.[†] The long-term ecosystem
benefits of managing agricultural land to enhance carbon sequestration are
arguably greater than the short-term increases in food production. Managing
land to increase carbon rather than depleting it means that soil resources,
including a full range of essential nutrients, are preserved and improved for
future use. Soils with adequate carbon stores help to mitigate global warming by keeping carbon out of the atmosphere and improve water quality by
reducing pollution from erosion. Soils with high carbon stores also retain
moisture better than depleted soils and thus are more resistant and resilient
to drought.

3.3.2 Cropland, Grazing Land, Water, Population, and Food

There are currently about 15–15.6 million square kilometers (5.79–6.02 million square miles) of cropland on the Earth.[‡] The range of estimates is due
to the complexity of accurate measurement using satellite technology and
continual changes in land use. Cultivated land is about 12% of the ice-free
land surface of the planet. The United Nations (UN) Food and Agriculture
Organisation (FAO) recently calculated that there is 0.23 hectares (0.57 acres)
of cultivated land per person globally, with only 0.17 hectares per capita in
low-income countries and more than twice that amount, 0.37 hectares per

---

[*] Lal et al., "Soil carbon sequestration to mitigate climate change."

[†] Rattan Lal, "Managing soils and ecosystems for mitigating anthropogenic carbon emissions
and advancing global food security," *BioScience* 60, no. 9 (2010).

[‡] Navin Ramankutty et al., "Farming the planet: 1. Geographic distribution of global agricul-
tural lands in the year 2000," *Global Biogeochemical Cycles* 22, no. 1 (2008); Food and Agriculture
Organisation, *The State of the World's Land and Water Resources for Food and Agriculture: Summary
Report* (Rome: United Nations, 2011).

capita, in high-income countries. That calculation was based on slightly fewer than 7 billion people in 2009.*

The extent of global grazing land is even more difficult to measure precisely than is cropland. Land managed as pasture, mixed grassland, and woodland ecosystems and sparsely vegetated or barren land are all sometimes used for grazing. Using the same satellite technology and methods used to identify 15 million square kilometers of cropland, Ramankutty et al. identified 28 million square kilometers (10.8 million square miles) of pasture. This is about 22% of the ice-free land surface.† Lal has recently estimated the global extent of grazing land at 34.5 million square kilometers (13.3 million square miles).‡ Combining cropland and pasture land, total agricultural land is at least 43 million square kilometers (16.6 million square miles). This total does not include many orchards or other perennial tree and bush crops and probably some marginal land used for grazing. In total, about 40% of the Earth's land surface is what we collectively have to work with to feed ourselves. This has already proven to be difficult, as food scarcity, malnutrition, and food insecurity have increased in recent years.§ The greater challenge clearly lies in the future.

World population reached 7 billion in 2011 and the midlevel projection by the UN is for the world's population to reach 9 billion by 2050. This population increase of about 29% in less than forty years, in combination with (assumed) rising incomes and changing food preferences (more meat, eggs, and grain), is projected by the UN to create a needed increase in food production of 70% globally and an increase in production of 100% in developing nations. That translates into a need for 1 billion additional metric tons of grain and 200 million metric tons of additional animal products by 2050. In 2011 the UN estimated that about 1 billion people, primarily in Asia (578 million) and sub-Saharan Africa (329 million) were undernourished.¶ Others place the current number of malnourished at 2 billion.** The UN estimates that 370 million people will remain at risk of being undernourished in 2050 even if their goal of a 70% increase in global food production and a 100% increase in developing nations is realized.††

The global cultivated land area has grown by only 12% over the past fifty years (1961–2011) while the total irrigated area has more than doubled in the same period. Agricultural production has grown between 250% and 300% since 1961, an increase due to large increases in the yield of major crops. These increases were made possible primarily by input-intensive fossil

---

* Food and Agriculture Organisation, *The State of the World's Land and Water Resources*.
† Ramankutty et al., "Farming the planet: 1."
‡ Lal, "Managing soils and ecosystems for mitigating anthropogenic carbon."
§ Lester Russell Brown, *World on the Edge: How to Prevent Environmental and Economic Collapse*, 1st ed. (New York: W. W. Norton, 2011).
¶ Food and Agriculture Organisation, *The State of the World's Land and Water Resources*.
** Tim Jackson, *Prosperity Without Growth: Economics for a Finite Planet* (London: Earthscan, 2011).
†† Food and Agriculture Organisation, *The State of the World's Land and Water Resources*.

fuel-based fertilizers, mechanization and the expansion of irrigation. Nearly half of the increase in food production during the past fifty years came from irrigated areas.[*] About 40% of current food production comes from the 2% of global cropland that is irrigated.[†]

The imperative for agricultural production growth through 2050 identified by the FAO has very high stakes and includes some very problematic assumptions about the continued availability of sufficient water and soil resources. The consequences of failure to meet increased agriculture production targets, in the context of adding 2 billion people to the global population, include widespread malnutrition (at best) and deadly famine for hundreds of millions or billions of people (at worst). Unlike conventional goals for general economic growth, agricultural productivity growth requires a larger material output of food or people will starve. There is no substitute for soil or water, and the demand for increased productivity is coming at the end of a long period of soil degradation and declining water availability. The FAO recognizes the risks associated with these resource scarcities and acknowledges: "The challenge of providing sufficient food for everyone worldwide has never been greater."[‡]

The FAO also recognizes that the rate of growth in global agricultural production has already begun a precipitous decline; the rate of production growth in developing countries is now half of the 3% annual growth rates seen in the past. Several recently emerging changes are exacerbating the effects of long-term depletion and degradation of water and soil resources. The food price shocks of 2007 and 2008 were precipitated by soaring grain prices that were in turn triggered by a Russian crop failure due to widespread drought and increasing production of grain and soy-based biofuels.[§] As global demand moves closer to annual production levels, shrinking grain stores will provide a diminished cushion against supply bottlenecks and shortages and accompanying price spikes. Those nations and individuals with the fewest resources in terms of land and water resources and in terms of income and wealth will continue to be at the greatest risk of food insecurity.[¶]

Increased competition for water and land is coming from many fronts. Sovereign and corporate investors have begun to acquire large tracts of land in developing countries and are thus reducing access to land and water by local producers. The rapid global expansion in production of feedstocks for biofuels on prime cultivated land is competing with food production. Climate change is increasing the risk and unpredictability for farmers because of warming and related aridity and from general changes in rainfall patterns. Increases in the frequencies of floods,

---

[*] Food and Agriculture Organisation, *The State of the World's Land and Water Resources*.
[†] Brown, *World on the Edge*.
[‡] Food and Agriculture Organisation, *The State of the World's Land and Water Resources*, 4.
[§] Brown, *World on the Edge*.
[¶] Jackson, *Prosperity Without Growth*.

droughts, and landslides further undermine the stability and reliability of food production on established croplands.[*] Sea level rise threatens river deltas and coastal areas that produce much of the world's rice, especially in Southeast Asia.[†]

The FAO, like others who have confronted the thorny dilemma posed by food production needs, population growth, and depleted land and water resources, recognizes that future expansion in the extent of cultivated land will be very limited because the vast majority of suitable land is already in use. Limited expansion of cultivated land may be possible in sub-Saharan Africa, Latin America, and in northern temperate areas as global warming expands the northern range of the grain-producing regions. Land under irrigation is projected to expand by only 6%, from 301 million hectares in 2009 to 318 million hectares (744 million acres to 786 million acres) in 2050. Recognizing that the expansion of irrigated and rain-fed cropland is severely constrained by a lack of available land and water resources, the FAO identifies "intensification of production on existing agricultural land" as the primary source of most future increases in agricultural production.[‡] Conversion of grassland or forest has been the historical source for expanding cropland. Between 1700 and 1900 about 1.4 billion hectares (3.5 billion acres) of forest and woodlands were converted to croplands and about 770 million hectares (1.9 billion acres) of savanna, grassland, and steppe were converted to cropland.[§] Most of the land suitable for cropland, and much that is not, has already been cleared and cultivated. The expansion of agricultural lands has been and remains the primary cause of deforestation, terrestrial carbon emissions, soil degradation, and biodiversity loss.

Competition for land and water resources between local small landholders and large corporations and nation-states searching worldwide for additional productive capacity will likely intensify inequality between those with capital and the poor. The subsidized expansion of grain-based biofuels at the expense of food production has not yet run its course despite accumulating evidence that grain-based biofuels make little ecological sense except as a bridge technology to next-generation biofuels from cellulose.[¶] Climate change is having a growing impact on agriculture due to changing rainfall patterns, extreme weather events, sea level rise, warming temperatures, aridity, and greater uncertainty. All of these trends magnify the most profound material constraint on continued growth in agricultural production: long-term degradation and depletion of land and water resources. Ecological limits, in this case scarcity of nonsubstitutable

---

[*] Food and Agriculture Organisation, *The State of the World's Land and Water Resources*.
[†] Brown, *World on the Edge*.
[‡] Food and Agriculture Organisation, *The State of the World's Land and Water Resources*, 11.
[§] Lal, "Managing soils and ecosystems for mitigating anthropogenic carbon."
[¶] Rachel Ehrenberg, "The biofuel future: scientists seek ways to make green energy pay off," *Science News* 176, no. 3 (2009).

natural resources, are emerging as a physical constraint on the continued growth of food production.

Water demand has tripled over the past fifty years and 70% of global water use is for irrigation. Water scarcity is growing on every continent. Many of the world's largest rivers are flowing at less than 5% of their former water volume. Rivers, streams, lakes, and aquifers are polluted with chemicals, nutrients, and salts. Large lakes and reservoirs are shrinking from lower stream flows and filling with sediment from eroding soils. One-half of the wetlands in North America and Europe are gone. The global distribution of physical water scarcity now includes most of northern and central Asia and the Middle East, India, the southwestern United States, northern Mexico, the west coast of South America, southern Australia, and about one-third of Africa, mostly in the north, east, and south.[*]

Saudi Arabia has depleted the aquifer that once allowed it to grow the majority of its grain and the Saudis are now importing almost all of their grain and buying and leasing land in Ethiopia and Sudan to gain access to land and water resources. Yemen is rapidly depleting its aquifers and now imports more than 80% of its grain. Unlike neighboring Saudi Arabia, Yemen's oil resources are also rapidly declining and the Yemenis cannot afford to purchase African or Latin American resources. The majority of Yemeni children are chronically undernourished; the state is in danger of disintegration and devolving into tribal wars over remaining water resources.[†] This is only one example of the social, political, and ecological implications of water scarcity that is rapidly spreading.

About half of the world's population lives in nations where water is being pumped from aquifers faster than it can be recharged by precipitation or where aquifers contain "fossil water" that is not recharged to a significant degree. These scarcities are not limited to relatively small countries like Saudi Arabia or Yemen; water tables are falling and wells are going dry in key grain-producing countries, including the United States, India, and China, which together produce about half of the world's grain. The Ogallala Aquifer underlying parts of eight states in the U.S. Great Plains and the deep aquifer under the North China Plain are both fossil aquifers and both are supplying irrigation water to major grain-producing regions.[‡]

Global groundwater resources have been monitored since 2002 by a pair of gravity monitoring satellites. They are part of a joint project by the National Aeronautics and Space Administration (NASA) and the German Aerospace Center known as the Gravity Recovery and Climate Experiment (GRACE). The GRACE satellites take monthly readings of global aquifers and can detect changes in groundwater levels down to 1 centimeter. Jay

---

[*] Food and Agriculture Organisation, *The State of the World's Land and Water Resources*.
[†] Brown, *World on the Edge*.
[‡] Ibid.

Famiglietti, a hydrologist with the project in Irving, California, provided a succinct summary of the GRACE findings: "groundwater is being depleted at a rapid clip in virtually all of the major aquifers in the world's arid and semiarid regions."* The decline in groundwater levels is most pronounced in areas where expanding agriculture has increased demand for water. Parts of China, India, the Middle East, and California are showing rapid declines and groundwater is disappearing entirely beneath southern Argentina, western Australia, and parts of the western United States. The GRACE satellites can measure changes in water levels but cannot provide accurate measures of the amount of water remaining in aquifers. In many parts of the world there are no reliable estimates of the amount of water stored in aquifers; we know major aquifers are being pumped faster than they recharge, but in many cases it is not known how long the water will last.†

Powerful pumps driven by electric or diesel motors have made it possible for irrigators to remove water from aquifers at rates that exceed recharge. When water is pumped from aquifers that have little or no natural recharge or water is pumped at a rate that exceeds recharge, the water level falls and the water must be pumped from greater depths. This results in water eventually becoming unavailable in sufficient quantities to support irrigation in those areas above the shallower portions (known as low saturated thickness) of the aquifer. This has been occurring in the Ogallala Aquifer in the panhandle of Texas and in parts of eastern Colorado and western Kansas. Aquifer depletion and diversion of irrigation water to fast-growing cities has resulted in shrinking irrigated areas in California, Texas, Arizona, Colorado, and Florida. It is likely that the irrigated area in the United States has peaked.‡

Despite a long history of evolving regulatory strategies and improved technological efficiencies, the rate of extraction of Ogallala water in my home state of Kansas continues to deplete the aquifer. Artificial recharge of the fossil aquifer has recently been offered as a strategy for slowing the rate of depletion.§ It is not surprising that desperate and ecologically risky technological fixes and a range of regulatory schemes are under consideration to slow the depletion of the Ogallala. Irrigated corn, cattle feedlots, beef processing plants, and coal-fired power plants in the region are dependent on this declining water resource. Largely eliminating all four of these industrial giants may prove to be necessary when the water has been sufficiently

---

* Devin Powell, "Groundwater dropping globally: satellites find supply falling mostly due to agriculture " *Science News* 188, no. 1 (2011): 5, http://www.sciencenews.org/view/generic/id/337097/title/Groundwater_dropping_globally

† Powell, "Groundwater dropping globally," 5.

‡ Brown, *World on the Edge.*

§ Sophocleous Marios, "The evolution of groundwater management paradigms in Kansas and possible new steps towards water sustainability," *Journal of Hydrology* 414–415 (11 January 2012).

depleted. Similar situations are emerging, at different rates and magnitudes, around the globe.

Surface waters are disappearing as well. Snow pack volume is declining and melting earlier in most mountainous regions and glaciers are melting at accelerating rates as warming temperatures are more pronounced at high elevations. Central Asia and the Tibetan Plateau contain glaciers with the largest extent of ice outside the Arctic and Antarctica. This "third pole" region is one of the world's most rapidly warming areas and the rate of glacial melting is now being closely monitored with a new network of high-altitude weather stations. The glaciers of central Asia and the Tibetan Plateau feed water to the Indus, Brahmaputra, Yangtze, and several other major Asian rivers. More than 100 million people depend on this threatened water source.[*]

Water scarcity is the product of a wide range of human actions, from over-pumping aquifers to filling lakes and reservoirs with eroded soils, to burning fossil fuels. At least some of the scarcity created by human actions can be addressed by reducing and reversing damaging practices. No substitute for water exists. Improving the productivity of a limited water supply needed for a growing population will require reversing and replacing a range of overly consumptive and polluting water management practices. This is a key example of the need to shift away from the existing industrial economic goal of improving labor productivity to a new emphasis on enhancing resource productivity.[†] Labor is abundant; soil and water are scarce.

Soil erosion, declines in soil organic carbon, and low soil fertility are even more pervasive in their effects on global agricultural lands than water scarcity. As noted above, Rattan Lal estimates that the cumulative soil organic carbon loss in global agroecosystems soils is 25%–75%, depending on climate, soil type, and management practices. This amounts to a loss in the range of 10–50 tons of carbon per hectare (4–20.2 tons per acre) since agricultural soil disturbance began. Lal notes that soils with severe depletion of soil organic carbon have low productivity and low use efficiency of inputs of fertilizer and water.[‡] Low levels of soil organic carbon also intensify the effects of drought by reducing soil water retention.[§]

The annual dust storms that originate in China and arrive in South Korea in late winter and early spring are known by Koreans as "the fifth season." The number of days that Koreans must contend with these dust events has been increasing since the 1970s. The Chinese dust storms occasionally reach as far as the United States, as they did in both 2001 and 2010, and serve as a dramatic indicator of the soil erosion that has been accelerating since the

---

[*] Christina Larson, "'Third Pole' glacier research gets a boost from China," *Science* 334, no. 6060 (2011).

[†] Jackson, *Prosperity Without Growth*.

[‡] Lal, "Sequestering carbon in soils of agro-ecosystems."

[§] R. Lal, "Soils and food sufficiency. A review," *Agronomy for Sustainable Development* 29, no. 1 (2009).

beginning of agriculture.* Desertification is perhaps the most extreme form of erosion and now affects 25% of the Earth's land area. When vegetation is removed by excessive tillage, overgrazing, aridification, or a combination of these factors, wind blows small soil particles away first and then the larger sand particles. Dust storms give way to sand storms and create deserts with little or no productive capacity remaining. The 1930s Dust Bowl in the Great Plains of the United States was primarily caused by overplowing and an extended drought. The widespread desertification occurring now in western and northern China is due mostly to overgrazing. China's 92 million cattle and 291 million sheep and goats are so concentrated in these regions that they are removing the protective vegetation from the land. The desertification in China extends into Mongolia and large areas of central Asia. Another large area plagued by dust storms, sand storms, and desertification is growing in the Sahel region of central Africa, to the south of the Sahara Desert. In both cases overgrazing, overplowing, deforestation, and drought/aridification are combining to create the largest conversions of agricultural land to desert in history.†

We saw in Chapter 1 that the International Commission on Stratigraphy, in considering some of the most significant and lasting human impacts on the geology of the planet, noted that more than 90% of the Earth's vertebrate biomass is now made up of humans and domesticated animals.‡ The impacts of rapidly growing populations of grazing animals, especially cattle, sheep, and goats, are not limited to the mark they will leave on the fossil record. The expansion of grazing lands for cattle is the largest single proximate cause of tropical deforestation.§ The growth of cattle, and especially goat and sheep, populations in China, other central Asian nations, and much of Africa is a central driving force in grassland degradation and desertification. China lost about 1,538 square kilometers (600 square miles) per year to desertification from 1950 to 1975. Since 1987 the loss to desertification has averaged 3,564 square kilometers (1,390 square miles) per year. This increasing rate paralleled the exponential growth of grazing animal populations. The Chinese cattle herd is comparable to that of the United States, at about 92 million head. The population of goats and sheep in China, however, is 281 million, compared to only about 9 million in the United States. It is no coincidence that these animals are concentrated in the northern and western Chinese provinces that are turning to desert.¶

A similar process is occurring throughout much of Africa, where the human population increased from about 227 million in 1950 to 1 billion in 2009 and livestock numbers increased from about 300 million to 862 million. Grassland carrying capacity is now often exceeded by half or more. In

---

* Brown, *World on the Edge.*
† Ibid.
‡ Gaia Vince, "An epoch debate," *Science* 334, no. 6052 (2011).
§ Geist and Lambin, "Proximate causes and underlying driving forces of tropical deforestation."
¶ Brown, *World on the Edge.*

Nigeria, desertification is claiming 351,000 hectares (867,000 acres) of range-land and cropland per year. Nigeria's population increased from 37 million in 1950 to 151 million in 2008, while its livestock population grew from 6 million to 104 million. The previously productive grasslands in northern Nigeria are turning to desert as 16 million cattle and 88 million sheep and goats greatly exceed the carrying capacity of the pastures.[*]

The explosion of the global goat population in recent decades is an indicator of the poor health of grasslands. Global cattle numbers increased by 28% between 1970 and 2009, the sheep population was stable during this period, while the goat population more than doubled.[†] As grasslands are degraded by overgrazing, grass is often replaced by woody shrubs. Cattle and sheep are well suited for grass and do not do well if forced to browse on shrubs. Goats are well adapted to the coarse forage on degraded landscapes and may be viewed as the only viable option by ranchers and herders with access to grazing lands that have been perennially overgrazed. Restoring organic carbon and nutrients to overgrazed grasslands will require reducing the number of grazing animals in some ecosystems. Careful management of the timing and density of grazing in conjunction with planting or restoring multiple species of perennial grasses in combination with legumes can build soil, increase ecological diversity and increase resistance and resilience to droughts and flooding. These changes will also enhance primary production and carrying capacity, but only if the degradation caused by overgrazing is reduced.

3.3.3 Wetlands, Peatlands, and Aquatic Ecosystems

Inland water systems and wetlands are important components of the terrestrial carbon cycle and terrestrial ecosystems. Soil organic carbon and nutrients are both deposited and removed by flowing water. Surface water is widely used for irrigation and terrestrial aquatic ecosystems are sinks for large stores of carbon. Peatlands are composed of deep soils with a minimum of 65% organic matter that forms in the anaerobic (low oxygen) conditions associated with wetland areas. Wetland habitats like marsh and bog areas are vulnerable to destruction when they are drained and claimed for agriculture, forestry, pest control, or a variety of other human purposes.[‡] Peatlands store about 400 gigatons of carbon and have contributed to cooling the climate for thousands of years.[§] To put this in perspective, 400 gigatons of carbon is slightly more than the carbon stored in boreal (272 gigatons) and temperate (119 gigatons) forests combined.[¶] Many peatlands are located in forested and agricultural areas, so they cannot be considered an entirely

---

[*] Brown, *World on the Edge.*
[†] Ibid.
[‡] Michael Allaby, *A Dictionary of Ecology*, 3rd ed. (Oxford: Oxford University Press, 2005).
[§] Lal, "Managing soils and ecosystems for mitigating anthropogenic carbon."
[¶] Pan et al., "A large and persistent carbon sink in the world's forests."

separate and additional carbon sink. Wetlands are a particularly carbon-rich component of forested and agricultural lands and require special consideration and protection.

3.4 Food Security

Is it possible to increase global food production by 70%, with a 100% increase of food production in low-income developing countries, by 2050? This is the ambitious goal set by the FAO to (nearly) meet the nutritional needs of 9 billion people. After an extensive review of global agricultural production statistics and recent evidence on the extent and rate of degradation of global land and water resources, the FAO offers a tentative and qualified "maybe." *If* the farmers of the world can implement the required intensification of production on existing agricultural land by making effective and efficient use of water and land resources to a degree that is unprecedented in the history of agriculture, *and* do so without causing further harm to land and water resources, then maybe the goal can be met. The rate of ongoing global warming, sea level rise, and increased frequency of droughts, floods, and extreme weather events are uncertain and considered by the FAO to likely provide significant challenges beyond those posed by direct resource scarcities. Acquisition and accumulation of land and water resources in low-income nations by foreign states and corporations with capitol resources is likely to make any increase in food production in developing nations less available to those who need it most.[*]

There are several implicit assumptions underlying the FAO's otherwise informed and reasonable assessment of population and global land and water resources that deserve explicit consideration. The midlevel UN projection of a human population of 9 billion by 2050 requires the continuation of recent population trends. An important component of those trends is the declining population growth rates recently achieved in many developing nations, primarily due to declining poverty rates and improved economic and education opportunities for girls and women. The improvements made in reducing global poverty from about 1990 to 2005, however, have stalled. In 2009 the World Bank reported that the incidence of poverty increased from 2005 to 2008 in east Asia, the Middle East, south Asia, and sub-Saharan Africa, with an increase of at least 130 million people classified as extremely poor, that is, living on less than $1.25 per day.[†] The high food prices of 2008 and the global economic crisis that emerged at the same time may well be the new normal and the end of an era of poverty reduction. Higher rates

---

[*] Food and Agriculture Organisation, *The State of the World's Land and Water Resources*.
[†] L. R. Brown, Plan B: *Mobilizing to Save Civilization* (Washington, DC: Screenscope, 2010).

of extreme poverty and illiteracy may mean a return to higher population growth rates in the short term.

If higher food prices, climate change, and resource scarcities accelerate and lead to widespread famine, we may never reach a population of 9 billion. This appears to be the more likely scenario. The UN estimate of 9 billion by 2050 and more than 10 billion by 2100 assumes a continuation of recent trends into discontinuous times. Population forecasts by the UN and by most demographers tend to largely ignore economic and resource constraints. Global warming and the ecological changes that accompany it—desertification, water shortages, loss of biological diversity, and general degradation of natural resources—could have catastrophic effects on population. Despite abundant and accumulating evidence of profound global environmental change, mainstream demographers have not yet accounted for these changes in their projections of population growth, which continue to be accompanied by assumptions of continued economic growth.*

The second set of assumptions made by the FAO is related to the first and involves the assumed continuation of trends in developing nations toward higher incomes and changing diets that include increased consumption of animal products and grains. Continuation of these trends would require continued economic growth or a redistribution of wealth and income to the poor, either from the global north, from economic elites in developing nations, or both. With the exception (for now) of China, the return of increasing global poverty and food insecurity in recent years signals the precarious and unsustainable nature of the post-1990 period of widespread economic growth experienced throughout much of Asia and Africa.

Assuming continued economic growth in the global economy through 2050 requires faith in some rapidly emerging technological fix to resource scarcity or a delusional avoidance of the reality of overloaded ecological sinks, including greenhouse gases in the global atmosphere and acidification of the oceans. As long as economic growth involves increasing the throughput of energy and material resources, its indefinite continuation is literally impossible. The eventual end of economic growth may reduce the needed expansion of global food production as both population growth and per capita consumption decline. If the end of overall economic growth is accompanied by a redistribution of wealth and food to those in need, poverty and malnutrition could decline as well. These would be the essential features of a postgrowth equitable and ecologically sound civilization that is organized to meet the needs of all.† Making the transition to such a civilization will

---

* Ronald Lee, "The outlook for population growth," *Science* 333, no. 6042 (2011).
† Herman E. Daly and Joshua C. Farley, *Ecological Economics: Principles and Applications*, 2nd ed. (Washington, DC: Island Press, 2010); Greer, *The Wealth of Nature*; Heinberg, *The End of Growth*; Jackson, *Prosperity Without Growth*.

require the collective and democratically coordinated efforts of individuals, communities, organizations, and governments on local and global scales.

An essential step in the transition to a more ecologically sound civilization is to change our relationship with and care of the land that surrounds us and upon which we are utterly dependent. Polanyi recognized more than half a century ago that the appropriation and commodification of land for the purposes and exploitive use of an industrial market economy represented an existential separation of humans from their social and material roots. Appropriation of the land and water resources of people in poor nations by those with capital continues today. Access to land and water to produce food and provide shelter is the most fundamental of needs for billions of people. Most of us in wealthy industrial nations are separated from a direct relationship to land as the basis for survival. We pay others to grow, process, transport, and, increasingly, cook and serve our food. Despite this illusory separation we are all as dependent on the Earth's finite store of soil and water resources as the poorest farmer or herdsman.

The threat posed by the wealthy hoarding resources in poor nations pales in comparison to the threat manifested in the degradation and depletion of soil and water resources that are continuing today nearly everywhere crops are grown or domestic animals are grazed. The FAO estimates that only 10% of all agricultural land is managed in a way that it is improving in quality, 18% is bare of vegetation, 25% is highly degraded and degradation is continuing at a high rate, 8% is moderately degraded and degradation is continuing at a moderate rate, and 36% is slightly to moderately degraded and in a stable condition. Water for irrigation is in "physical scarcity" in much of the world, including some of the most productive grain-producing regions.[*] Despite these trends the FAO projection for a 70% increase in food production by 2050 is based on the requirement that management of land and water resources must "improve markedly" to "reverse trends in their degradation."[†]

Food security in the future unequivocally depends on halting and reversing what has been an accelerating 10,000 year process of soil degradation and water resource depletion. Immediate, safe, and cost-effective mitigation and adaptation to anthropogenic climate change requires, in part, the very same reversal. In a confluence of global ecological proportions, the very same land management practices required for food security and for an essential component of climate change mitigation and adaptation will also make an essential contribution to the preservation and restoration of intact ecosystems and the enhancement of biological diversity. The management and care of forests, croplands, grasslands, wetlands, and the rest of the terrestrial system to enhance carbon uptake and sequestration can make a major contribution to each of these aspects of needed ecological restoration.[‡]

---

[*] Food and Agriculture Organisation, *The State of the World's Land and Water Resources*.
[†] Food and Agriculture Organisation, *The State of the World's Land and Water Resources*, 12.
[‡] Lal, "Managing soils and ecosystems for mitigating anthropogenic carbon."

3.5 Beyond Emissions

The fortunate confluence of positive ecological benefits associated with enhanced terrestrial carbon biosequestration is, as noted earlier, underappreciated and underemphasized in many discussions of climate change, ecological decline, and food security. Reducing emissions from fossil fuels, the source of more than 70% of current greenhouse gas emissions is absolutely essential to addressing climate change and many of the ecological problems that define the Anthropocene epoch. Even the total elimination of fossil fuels, however, will not restore the carbon lost from agricultural soils that must be replenished to preserve and increase crop and photosynthetic productivity. Forests and wetlands perform essential ecological services for humans and millions of other species. Restoration of these ecosystems necessarily involves the replenishment and protection of organic carbon in plants and soils. The elimination of fossil fuels will not directly or sufficiently protect scarce water resources or preserve the essential ecological diversity of intact ecosystems. The elimination of fossil fuels for all but the most limited and essential purposes is necessary but not sufficient to allow our descendants a fair chance for a healthy and prosperous future. Enhancing carbon biosequestration in terrestrial ecosystems is also essential.

It is possible for nearly everyone to make a contribution to enhanced carbon sequestration. An apartment dweller can make compost from kitchen wastes and use the carbon-, humus-, and nutrient-rich product to grow food or other plants in pots on a patio or in a window planter. The city or suburban resident with a small lot can compost leaves and grass clippings (or leave them where they fall), plant trees and other perennial plants, and keep garden soil covered with mulch. Farmers can plant trees in riparian areas and on field borders. Trees or perennial grasses can be planted on marginal sloping cropland that is highly erodible. Cropland can be improved with cover crops, legumes, and well-planned crop rotations. Tillage can be eliminated in some cases and minimized in others. The productivity and carbon content of pasture can be improved with close monitoring of stocking rates, using carefully timed paddock rotations, planting or maintaining deep-rooted perennial species, and interseeding with legumes.

Agroforestry systems, which intersperse tree crops with field crops and have been utilized in many parts of the tropics for thousands of years, can be renewed and expanded to minimize soil degradation from exposure to direct sunlight and heavy seasonal rains. Perennial tree and woody shrub crops can be planted to replace some annual crops that require frequent cultivation and soil disturbance, especially on sloping land of marginal productivity that is prone to erosion. Plant breeders, ecologists, and other scientists are working to develop perennial grains that can be grown in polycultures and may, over the next fifty years, replace most annual grain crops grown in monocultures.*

All of these existing and emerging land management practices, along with afforestation, reforestation, restoration of degraded forests and wetlands, and protection of intact forest and wetland ecosystems have the capacity to enhance terrestrial carbon sequestration and protect and conserve scarce water resources. They all protect soil and water resources from degradation by building soil carbon, soil nutrients, and total biomass. Widespread adoption of these and similar practices will make a positive contribution to the mitigation and adaptation to anthropogenic climate change, enhance future food security for a growing population, and help to protect remaining ecological diversity.

3.6 A Conservative Estimate of Global Terrestrial Carbon Biosequestration Enhancement

It is now possible to make an estimate of the total net carbon sequestration enhancement potential for the Earth's terrestrial ecosystems over the next few decades. I use the word "potential" not in the sense of maximum biological potential, which has been estimated at as much as 5.65–8.71 gigatons per year with no social or economic constraints[†] and at 2.55–4.96 gigatons per year by Lal,[‡] but rather in terms of a reasonable and realistic midpoint estimate based on the evidence summarized here.

My assumptions regarding forests are as follows: (1) boreal forests can be managed so they continue to sequester about 0.5 gigatons of net carbon per year; (2) the recent increase in net carbon sequestration in temperate forests will continue due to expanded forest area over the next decade at the approximate rate of increase seen from the 1990s to the 2000s and thus will increase from 0.8 gigatons of carbon per year to 0.9 gigatons of carbon per year; (3) tropical deforestation can be slowed by 25% in a decade, which is similar to the slowing rate of tropical deforestation over the past decade—this will reduce gross tropical deforestation carbon emissions from 2.8 gigatons to 2.1 gigatons, a decrease in emissions of 0.7 gigatons. These changes result in a net increase in annual global forest carbon sequestration by 2022 of 0.8 gigatons. These gains can be achieved by maintaining the direction and rate of change achieved from the 1990s through the 2000s.[§]

---

[*] Jackson, *Consulting the Genius of the Place.*

[†] Pete Smith, "Engineered biological sinks on land," in *The Global Carbon Cycle: Integrating Humans, Climate, and the Natural World,* ed. Christopher B. Field and M. R. Raupach (Washington, DC: Island Press, 2004).

[‡] Lal, "Managing soils and ecosystems for mitigating anthropogenic carbon."

[§] Pan et al., "A large and persistent carbon sink in the world's forests."

Achieving an increase in annual forest carbon sequestration of this magnitude will require continued expansion of temperate forest biomes in areas not limited by aridification, careful protection and rapid replanting of boreal forests threatened by warming related fires and insect infestations, and continuing to slow agricultural and infrastructure expansion into tropical forests in Latin America, sub-Saharan Africa, and Asia. The estimate of 0.8 gigatons of additional carbon sequestration per year by 2022 in the Earth's forests would constitute an increase to 2.0 net gigatons per year, up from the current rate of about 1.2 net gigatons per year.[*] This estimate includes gains due to afforestation and forest plantations, reforestation, forest succession, agroforestry, and protection of peatlands in forest biomes.

For agricultural land the potential gains have been estimated by Lal at 1.2–3.1 gigatons per year of increased carbon sequestration through restoration of degraded and desertified soils and adoption of proven recommended practices on cropland and in grassland ecosystems.[†] The baseline data on increases or decreases in carbon sequestration that are available for global forests since 1990 are, unfortunately, not available for agricultural lands. Increases in soil carbon per hectare or acre associated with recommended practices such as reduced tillage, application of manure or compost, the use of crop residues and mulch, conversion to deep-rooted perennials, reduced stocking rates on pasture, and planting of cover crops and legumes have been measured in a variety of agricultural ecosystems. These measurements range from 100 kilograms per hectare per year to 1,000 kilograms per hectare per year (89–892 pounds per acre) depending on the soil type, degree of degradation and the practice applied.[‡] Applying the midrange estimate of 500 kilograms per hectare (446 pounds per acre) of carbon gains to the low-end estimate of 43 million square kilometers of global agricultural lands (4.3 billion hectares [10.63 billion acres]) would result in an additional sequestration of about 2.15 gigatons of carbon per year. Application of restorative land management practices on all agricultural lands can serve as a long-term goal, but is not a reasonable expectation by 2022. If carbon restoration practices were applied to a modest 25% of global agricultural lands by 2022, the increased carbon sequestration would be about 0.54 gigatons of carbon per year. Since about 10% of global agricultural lands are currently managed in ways that improve soil quality,[§] this would require a modest improvement in the proportion of global agricultural lands managed with best practices.

An additional 0.8 gigatons of carbon biosequestration per year in the Earth's forests and 0.54 gigatons of additional annual carbon biosequestration in global agricultural ecosystems would produce a total increase of 1.34 gigatons of carbon sequestered by the Earth's terrestrial ecosystems

[*] Pan et al., "A large and persistent carbon sink in the world's forests."
[†] Lal, "Sequestering carbon in soils of agro-ecosystems."
[‡] Lal et al., "Soil carbon sequestration to mitigate climate change."
[§] Food and Agriculture Organisation, *The State of the World's Land and Water Resources.*

per year by 2022. This is obviously not intended as a precise projection, but is a reasonable and conservative estimate based on the best information available. An estimated 1.34 gigaton per year increase in annual global terrestrial carbon biosequestration by 2022 is a reasonable goal using known and proven practices implemented on a modest scale. This level of increased biosequestration would decrease the net carbon currently accumulating in the atmosphere each year by about one-third if Pan's estimate of 4.1 gigatons of net atmospheric carbon accumulation is accurate.[*] A wider adoption of restorative land management on 50% of agricultural lands could potentially double the enhanced carbon sequestration in agricultural lands to more than 1 gigaton per year. If biochar proves to have the impact projected by the scientists researching its potential, another gigaton of carbon biosequestration per year could be realized.[†]

Of course, global carbon emissions from fossil fuels are continuing to accelerate. Any increase in terrestrial biosequestration of carbon will slow the rate of carbon accumulation in the atmosphere. Significant reductions in atmospheric carbon concentrations will also require first stabilizing and then reducing fossil fuel emissions. Both sides of the emissions/sequestration equation are essential and the terrestrial sequestration side is substantial, available immediately, and relatively inexpensive. Carbonizing terrestrial ecosystems will enhance food security and ecological diversity.

The potential to significantly enhance sequestration in agricultural ecosystems is due in large part to the long history of biomass and soil degradation from past and current agricultural practices. This presents a unique historical opportunity to recarbonize agricultural ecosystems to replace much of the carbon lost to the atmosphere over thousands of years.[‡] Terrestrial ecosystems have a finite capacity for biosequestration that may be limited to replacing the carbon cumulatively displaced from the land by human actions. That total has been estimated by Lal at about 320 gigatons of carbon over the past 6,000 years.[§]

Although the exact magnitude of recarbonization potential is not known it is clearly large in the context of the global carbon cycle. Working to enhance terrestrial carbon sequestration represents our best opportunity to immediately slow the accumulation of carbon in the atmosphere and thus buy time and adapt while we collectively develop the political will, technological infrastructure, and no-growth economic systems needed to reduce fossil fuel emissions to near zero.

---

[*] Pan et al., "A large and persistent carbon sink in the world's forests.

[†] Lal, "Managing soils and ecosystems for mitigating anthropogenic carbon." J. Lehmann, "A handful of carbon," *Nature* 447, no. 7141 (2007).

[‡] Ibid.

[§] Ibid.

References

Abramsky, K. 2009. *Sparking a Worldwide Energy Revolution: Social Struggles in the Transition to a Post-Petrol World*. Oakland, CA: AK Press.

Allaby, M. 2005. *A Dictionary of Ecology*, 3rd ed. Oxford: Oxford University Press.

Antonio, R. J. 2009. "Climate change, the resource crunch, and the global growth imperative." In *Current Perspectives in Social Theory*, vol. 26, ed. Harry F. Dahms, 3–73. Bingley, UK: Emerald Group Publishing.

Brown, L. R. 2010. *Plan B: Mobilizing to Save Civilization*. Washington, DC: Screenscope.

———. 2011. *World on the Edge: How to Prevent Environmental and Economic Collapse*, 1st ed. New York: W. W. Norton.

Chestney, N. 2012. "Global warming close to becoming irreversible-scientists." *Breaking US and International News*, http://www.reuters.com/assets/print?aid=USBRE82POUJ20120326

Daly, H. E., and J. C. Farley. 2010. *Ecological Economics: Principles and Applications*, 2nd ed. Washington, DC: Island Press

Dawson, T. P., S. T. Jackson, J. I. House, I. C. Prentice, and G. M. Mace. 2011. "Beyond predictions: biodiversity conservation in a changing climate." *Science* 332(6025):53–58.

DellaSala, D. A., A. Martin, R. Spivak, T. Schulke, B. Bird, M. Criley, C. van Daalen, J. Kreilick, R. Brown, and G. Aplet. 2003. "A citizen's call for ecological forest restoration: forest restoration principles and criteria." *Ecological Restoration* 21(1):14–23.

Ehrenberg, R. 2009. "The biofuel future: scientists seek ways to make green energy pay off." *Science News* 176(3):24–29.

Food and Agriculture Organisation. 2011. *State of the World's Forests*. Rome: United Nations.

———. 2011. *The State of the World's Land and Water Resources for Food and Agriculture: Summary Report*. Rome: United Nations.

Geist, H. J., and E. F. Lambin. 2002. "Proximate causes and underlying driving forces of tropical deforestation." *BioScience* 52(2):143–150.

Greer, J. M. 2011. *The Wealth of Nature: Economics as if Survival Mattered*. Gabriola, BC, Canada: New Society Publishers.

Heinberg, R. 2011. *The End of Growth: Adapting to Our New Economic Reality*. Gabriola, BC, Canada: New Society Publishers.

Intergovernmental Panel on Climate Change. 2007. *Climate Change 2007: Synthesis Report*, Fourth Assessment Report (AR4). Geneva: IPCC.

Jackson, T. 2011. *Prosperity Without Growth: Economics for a Finite Planet*. London: Earthscan.

Jackson, W. 2010. *Consulting the Genius of the Place: An Ecological Approach to a New Agriculture*. Berkeley, CA: Counterpoint Press.

Kiers, E. T., M. Duhamel, Y. Beesetty, J. A. Mensah, O. Franken, E. Verbruggen, C. R. Fellbaum, et al. 2011. "Reciprocal rewards stabilize cooperation in the mycorrhizal symbiosis." *Science* 333(6044):880–882.

Lal, R. 2008. "Carbon sequestration." *Philosophical Transactions of the Royal Society B: Biological Sciences* 363(1492):815–830.

———. 2009. "Soils and food sufficiency. A review." *Agronomy for Sustainable Development* 29(1):113–133.

———. 2010. "Managing soils and ecosystems for mitigating anthropogenic carbon emissions and advancing global food security." *BioScience* 60(9):708–721.

———. 2011. "Sequestering carbon in soils of agro-ecosystems." *Food Policy* 36(Suppl. 1):S33–S39.

Lal, R., R. F. Follett, B. A. Stewart, and J. M. Kimble. 2007. "Soil carbon sequestration to mitigate climate change and advance food security." *Soil Science* 172(12):943–956.

Larson, C. 2011. "'Third Pole' glacier research gets a boost from China." *Science* 334(6060):1199.

Lee, R. 2011. "The outlook for population growth." *Science* 333(6042):569–573.

Lehmann, J. 2007. "A handful of carbon." *Nature* 447(7141):143–144.

Marios, S. "The evolution of groundwater management paradigms in Kansas and possible new steps towards water sustainability." *Journal of Hydrology* 414–415 (11 January 2012): 550–559.

Maser, C. 1994. *Sustainable Forestry: Philosophy, Science, and Economics.* Boca Raton, FL: St. Lucie Press.

McDonough, W., and M. Braungart. 2002. *Cradle to Cradle: Remaking the Way We Make Things*, 1st ed. New York: North Point Press.

Nickens, T. E. 2009. "Paper chase." *Audubon*. 38–45.

Pan, Y., R. A. Birdsey, J. Fang, R. Houghton, P. E. Kauppi, W. A. Kurz, O. L. Phillips, et al. 2011. "A large and persistent carbon sink in the world's forests." *Science* 333(6045).

Polanyi, K. 1957. *The Great Transformation.* Boston: Beacon Press.

Powell, D. 2011. "Groundwater dropping globally: satellites find supply falling mostly due to agriculture," *Science News*, http://www.sciencenews.org/view/generic/id/337097/title/Groundwater_dropping_globally

Ramankutty, N., A. T. Evan, C. Monfreda, and J. A. Foley. 2008. "Farming the planet: 1. Geographic distribution of global agricultural lands in the year 2000." *Global Biogeochemical Cycles* 22(1):GB1003.

Seager, R., M. Ting, I. Held, Y. Kushnir, J. Lu, G. Vecchi, H.-P. Huang, et al. 2007. "Model projections of an imminent transition to a more arid climate in southwestern North America." *Science* 316(5828):1181–1184.

Smith, P. 2004. "Engineered biological sinks on land." In *The Global Carbon Cycle: Integrating Humans, Climate, and the Natural World*, ed. C. B. Field and M. R. Raupach. Washington, DC: Island Press.

Sundquist, E. T., K. V. Ackerman, L. Parker, and D. Huntzinger. 2009. "An introduction to global carbon cycle management." In *Carbon Sequestration and Its Role in the Global Carbon Cycle*, ed. B. J. McPherson and E. T. Sundquist. Washington, DC: American Geophysical Union.

van Mantgem, P. J., N. L. Stephenson, J. C. Byrne, L. D. Daniels, J. F. Franklin, P. Z. Fulé, M. E. Harmon, et al. 2009. "Widespread increase of tree mortality rates in the western United States." *Science* 323(5913):521–524.

Vidal, J. 2012. "World's giant trees are dying off rapidly, studies show." *The Guardian*, http://www.guardian.co.uk/environment/2012/jan/26/giant-trees-dying

Vince, G. 2011. "An epoch debate." *Science* 334(6052):32–37.

von Braun, J., and R. S. Meinzen-Dick. 2009. "'Land grabbing' by foreign investors in developing countries: risks and opportunities." IFPRI Policy Brief 13. Washington, DC: International Food Policy Institute.

4

Land Management Examples, Practices, and Principles

4.1 Land Management for Carbon Biosequestration and Ecological Diversity

4.1.1 Canadian Boreal Forest

In 2001 my family and I took advantage of an opportunity to purchase 134 hectares (330 acres) of land in northwestern Ontario, Canada. In early 2012 we bought an additional 65 contiguous hectares (160 acres) and now own and care for 198 hectares (490 acres) of mixed boreal forest. The land is in the highly productive southern boreal forest located in the Superior–Saint Lawrence region, a few miles west of Lake Superior's Black Bay at about 48.8 degrees north latitude. The predominant tree species are trembling (quaking) aspen, balsam poplar, white (Canadian) spruce, Jack pine, northern white cedar, paper (white) birch, balsam fir, black spruce, black ash, and northern (eastern) white pine. There are also a few tamaracks and a wide range of shrubs, herbs, and mosses—a diverse mixture of boreal flora and fauna.

A ridge that alternates between steep cliffs, large boulders, and loose talus runs from the north boundary for 2.4 kilometers (1½ miles) southwest to near the southwest corner and divides the property into a high plateau on the west and a low area on the east. The high ground is dominated by a fifty- to sixty-year-old stand of Jack pine with some black spruce, white birch, and balsam fir. On the low side near the glacial ridge the dominant species are northern white cedar and white birch, which is typical in and around boulders and loose talus in northwestern Ontario.[*] The majority of the lowland is a mixture of trembling aspen and white spruce with pure and mixed stands of all the species listed above. The high ground is at an elevation of about 335 meters (1,100 feet) above sea level, totals approximately 73 hectares (180 acres), and has a substantial amount of exposed bedrock. The low ground ranges from 30 meters to 60 meters (100 to 200 feet) lower in elevation, totals about 117 hectares (290 acres), and has deeper soil, including several areas of peat-producing wetlands (an organic soil classified as Histosol that is at least 40 centimeters [16 inches] deep with at least 65% organic matter by dry

---

[*] R. A. Sims and K. A. Baldwin, *Landform features in northwestern Ontario*, COFDRA report 3312 (Ontario, Canada: Ontario Ministry of Natural Resources, 1991).

FIGURE 4.1 (See color insert.)
Mixed boreal forest with a talus ridge in the background.

weight[*]) with standing water at least part of the year. The peat-producing wetlands and waterlogged soils cover about 8 hectares (20 acres) of the lowland area. The remaining 8 hectares (20 acres) are taken by the cliffs, boulders, and talus along the ridge that divides the land.

This area of northwestern Ontario is known as "Canyon County" because of the steep cliffs that are a common landform feature and the proximity of the Ouimet Canyon Provincial Park. Our land is affectionately known as "The Hole" because of a deep, granite-walled glacial lake that is among the cliffs that divide the property between the upper elevations of thin rocky soils and the lower elevations of thicker soils and heavier vegetation.

The most recent glaciation, the Wisconsinan, covered most of northwestern Ontario until about 10,000 years before the present (BP). At about 9,500 years BP the several ice lobes of the Wisconsinan were retreating northward, allowing the massive glacial Lake Agassiz, which covered much of central Canada, to drain into Lake Minong, a glacial lake that was a precursor to Lake Superior. The Superior basin was inundated by more than a dozen torrential flows of water as more than 4,000 cubic kilometers of water from Lake Agassiz were released into the Superior basin by the northward migration of the Superior ice lobe. Following the northward retreat of the

---

[*] Michael Allaby, *A Dictionary of Ecology*, 4th ed. (Oxford: Oxford University Press, 2010).

FIGURE 4.2 (See color insert.)
Glacial Lake: "The Hole."

glacier and removal of its weight, there was a massive uplift, or isostatic rebound, that continues today. The uplift allowed the remainder of Lake Agassiz to drain into the Superior basin. It was not until about 5,500 years BP that Lake Superior reached its approximate current size and shape. The land west and north of what is now Lake Superior was under ice until about 10,000 years BP and then mostly under glacial lakes until about 5,500 years ago.[*] Many of those lakes remain today, of course, and constitute vast water and ecosystem resources.

This recent geological history makes the boreal forest of northwestern Ontario, and nearly all of Canada, very young in terms of elapsed time for postglacial ecological development. This is important for the ecological character of the place. Before undertaking the task of managing land it is useful and can be very satisfying to gain some knowledge about the history and ecology of the place. This seems obvious, but is too often largely ignored. As summarized by Canadian forest ecologists R. A. Sims and K. A. Baldwin, "the ecological character of a forest site is a function of the complex interaction of many biophysical conditions: climate, soil parent material, ecosystem function, topographic effects and elapsed developmental time are all important factors in defining the ecological uniqueness of a particular site."[†]

---

[*] Sims and Baldwin, *Landform features in northwestern Ontario."*
[†] Sims and Baldwin, *Landform features in northwestern Ontario,"* 2.

A primary management objective for this piece of boreal forest is to protect the site from agricultural and other "development" that would threaten the stability of the carbon stored in the live biomass, deadwood, litter, soils, and wetlands. Pan estimates that boreal forests average 239 metric tons of stored carbon per hectare (97.6 tons per acre).[*] This relatively healthy, diverse southern boreal forest with about 8 hectares (20 acres) of peat-producing wetlands probably has more carbon per hectare than average. Using Pan's estimate, these 198.3 hectares (490 acres) store at least 47,394 metric tons of carbon.

To put this in context it may be helpful to consider this boreal carbon sink in relation to the carbon emissions produced by driving a car. A car that gets 30 miles per gallon driven for a distance of 10,000 miles will burn about 330 gallons, approximately 1 ton, of gasoline. Burning a ton of gasoline produces about 3 tons of carbon dioxide (CO_2),[†] which is equivalent to 0.819 tons of carbon. The carbon stored on an average 198.3 hectares of boreal forest is roughly equal to the amount of carbon produced by burning about 19 million gallons of gasoline. (Gasoline weighs 6.3 pounds per gallon; 87%, or 5.5 pounds, is carbon.)[‡] One hectare of boreal forest stores as much carbon, 239 metric tons (2,204 pounds each), as is produced by burning about 95,774 gallons of gasoline. This is roughly equivalent to the carbon released to the atmosphere in 1 year by 288 cars driven 10,000 miles each at 30 miles per gallon. Of course, it takes centuries for boreal forests to accumulate the carbon they store through photosynthesis and biosequestration—all the more reason to prioritize protecting healthy, functioning boreal ecosystems.

A second and closely interrelated management objective is to preserve, and if possible enhance existing intact habitats for a wide range of wildlife, fungi, and plant species. The species richness and habitat diversity of this intact boreal forest ecosystem is greater than it would be under any alternative use. The full range of species—from amphibians and songbirds to black bears and moose and from ferns and mosses to shrubs and fungi—is needed for a healthy forest. The trees and the other organisms that constitute the ecological community are mutually dependent. The rare and endangered Kirtland's warbler, for example, relies almost exclusively on young stands of Jack pine for breeding areas.[§] An intact forest ecosystem cleans the air, stabilizes watersheds, and provides flood prevention and clean water for fisheries and downstream human uses. It also provides opportunities for a range

---

[*] Yude Pan et al., "A large and persistent carbon sink in the world's forests," *Science* 333, no. 6045 (2011).

[†] R. H. Socolow, "Can we bury global warming?," *Scientific American* 293, no. 1 (2005).

[‡] www.fueleconomy.gov, "How can a gallon of gasoline produce 20 pounds of carbon dioxide?," http://www.fueleconomy.gov/feg/co2.shtml

[§] Russel M. Burns and Barbara H. Honkala, *Silvics of North America*, vol. 1, *Conifers*; Agriculture Handbook 654 (Washington, DC: U.S. Forest Service, 1990); Russel M. Burns and Barbara H. Honkala, *Silvics of North America*, vol. 2, *Hardwoods*, Agriculture Handbook 654 (Washington, DC: U.S. Forest Service, 1990).

of recreational activities from fishing and hunting to blueberry picking and hiking, and has profound aesthetic value to some.

A third objective involves long-term production and use of wood products in ways that do not diminish the ecological integrity of the site nor compromise the quality of future harvests. In this place, that means encouraging the development of healthy, diverse stands of the longer-lived and commercially valuable tree species that were predominant on the site prior to a series of timber harvests over at least the past 100 years. These valuable species, especially northern white pine, white spruce, northern white cedar, and Jack pine, provide more secure and longer-term carbon sequestration and storage than shorter-lived, colonizing species such as trembling aspen and white birch. They are also relatively valuable timber species and can be selectively harvested to maintain species- and age-diverse stands. White birch and quaking aspen are pioneer species after fire and logging disturbance, short lived, and are generally replaced over time with conifers, especially on this site with white spruce and eastern white pine. Trembling aspen are primarily used for pulp and chipboard. These wood and paper products store carbon for much shorter periods than spruce and pine lumber used for construction or furniture. White birch is used for a variety of purposes, including firewood, but is of relatively moderate value for durable wood products and is also relatively short lived.

Jack pine (*Pinus banksiana*) grows well on thin soils over granites and metamorphosed rock, as are found on the upland part of this site, and also will do well over limestone, clay soils, loam, peat, and even over permafrost. The native range of Jack pine extends from the Northwest Territories to Nova Scotia and south into the northeastern and north-central United States. It is the most widely distributed pine and grows further north than any other pine in North America. Jack pine stands often originate after forest fires and have developed in pure, even-aged stands in vast areas of Canada. They are shade intolerant and are often succeeded, like other colonizing species, by more shade tolerant species like black spruce and white spruce. Jack pines' capacity to emerge quickly after fire, its ability to survive on dry, nutritionally poor sites, and the high nutrient levels found in the needles and litter it drops[*] may make it an even more important boreal species as climate change continues to make much of the boreal biome dryer and more prone to fire.

Jack pine is sometimes cut at forty–fifty years of age and used as pulpwood. At sixty–seventy years it is used for poles and saw timber. In Ontario, the highest mean annual growth is at about sixty years. Jack pine stands often begin to deteriorate after sixty–eighty years, depending on the quality of the site. Healthy, 185-year-old trees have been found in northern Minnesota, however, and the oldest tree reported, 230 years old, was located in Ontario north of Lake Superior and east of Lake Nipigon, about 100 miles

---

[*] Burns and Honkala, *Silvics of North America*, vol. 1, *Conifers*.

(160 kilometers) from the site described here. Jack pine at sixty years of age is commonly only 10–12 inches in diameter at breast height (ABH), but can grow to more than 28 inches ABH on favorable sites.[*]

Jack pines are longer lived than the deciduous colonizers in this area—trembling aspen, bigtooth aspen and white birch—and have a special capacity to survive on poor, dry sites. I have been told by neighbors and family members of the logger that cut many of the Jack pine from our land in the late 1940s and early 1950s that they were used for ties on the Canadian National Railway line that runs through the area. Those ties may still be storing much of the carbon they sequestered as they were growing from about 1880 through 1950, even if they were replaced with new ties and are now being used for landscaping and retaining walls. Durable wood products retain sequestered carbon for decades to centuries.

Eastern (northern) white pine (*Pinus strobus*) is one of the most valuable trees for timber in eastern North America. It is found in southern Canada from eastern Manitoba to Nova Scotia on the east coast and south to western Tennessee, western Kentucky, and Delaware. A variety of the species grows in the mountains of Guatemala and southern Mexico. It is the provincial tree of Ontario and is planted widely both inside and outside its native range. (I have planted them on my Kansas farm with mixed success; deer prefer their soft needles as winter browse.) Northern white pine is a fast-growing conifer and is often used for reforestation projects. White pine competes best on well-drained sandy soil of medium quality and will outproduce most other commercially valuable species on such sites. Hardwoods will generally dominate on more fertile sites.[†] Seedlings are shade tolerant, but I have observed that saplings more than 3 or 4 feet tall often need to be released by thinning adjacent trees to provide sufficient light or they will stagnate and die. This is especially the case when white pines are in a mixed stand with young trembling aspen, a fast-growing hardwood.

We contracted with an independent logger over a three-year period from 2003 through 2005 to harvest mature trembling aspen from about 28 hectares (70 acres) of the low-lying portion of our property. About 1,000 cords (a volume 4 feet × 4 feet × 8 feet) of quaking aspen (locally referred to as poplar) were cut and marketed locally, primarily for pulp and chipboard. About 100 cords each of white birch and white spruce were also cut, with the birch sold as firewood and the spruce as saw logs and some veneer. The primary objectives were to open a mostly closed canopy of trembling aspen and white birch to release young white spruce and white pine and to enhance the natural regeneration of those species. We also wanted to utilize the wood and generate income from the mature aspens, estimated to be forty–fifty years old and rapidly deteriorating and dying. The spruce was

---

[*] Burns and Honkala, *Silvics of North America*, vol. 1, *Conifers*.
[†] Ibid.

FIGURE 4.3 (See color insert.)
Young white spruce after a selective harvest of declining aspen and white birch.

harvested with the intent of taking only those trees that were overcrowded or damaged by the logging operation.

The success of this timber harvest was mixed. In retrospect, the objectives of releasing young pine and spruce may have been better accomplished by a lighter and more selective harvest of the aspen, an approach not favored by the logger. The very open canopy from a near clear-cut harvest in some areas led to rapid regeneration of fast-growing aspen that quickly closed a low canopy in some areas so that young pine and spruce were again unable to successfully compete. The primary management activity in recent years has been to systematically walk the harvested area and open crown space for young conifers by cutting aspen saplings that are blocking sunlight. This is difficult work in some areas because of the exceptionally thick and rapid growth of the aspen. Young aspen stands in cutover areas provide excellent browse for deer and moose, and cover for grouse and a wide range of other wildlife.

White spruce (*Picea glauca*) is the foremost pulpwood and most important and widely dispersed commercial tree species in Canada. It generally grows much larger than black spruce (*Picea mariana*) and grows well in a wide range of well-drained soils while black spruce tends to thrive in wet soils, poorly drained thin soils, and bogs. White spruce produces lumber that is light and straight-grained, ideal for both pulp and general construction. It has a range that extends along the northern range of trees from the coast of the Bering Sea in Alaska to the Atlantic Coast in the east and south into the north-central and northeastern United States. It is an exceptionally long-lived tree, making it important for the production of biomass and carbon sequestration throughout most of the Canadian boreal forest. Stunted trees nearly 1,000 years old have been identified above the Arctic Circle. On good sites that have been protected from fire, white spruce 250–300 years old are common. White spruce does not produce seed in quantity until about thirty years of age in most natural stands. Years of high seed production are generally spaced two–six years apart, but may occur up to ten–twelve years apart. Trees more than 30 meters (100 feet) tall and 60–90 centimeters (24–36 inches) in diameter are common on good sites throughout the range, and much larger trees up to 55 meters (180 feet) tall and 90–120 centimeters (36–48 inches) in diameter are found on excellent protected sites.* A key component of our boreal forest management plan is to assist in the regeneration, survival, and long life of this valuable, long-lived species.

Northern white cedar (*Thuja occidentalis*) is rot and termite resistant. It is used primarily for products that come in contact with soil and water. Fence posts and docks made of northern white cedar are long lasting and require no preservatives. It is used to make log cabins, lumber, poles, bowls, boats, pails, potato barrels, shingles, and siding. Its boughs are used in floral arrangements and can be used to distill cedar leaf oil, used in medicines and perfumes.† Northern white cedar was likely the first North American tree introduced into Europe. It is known to have been grown in Paris by about 1536. Tea made from its foliage and bark is high in vitamin C and was used as an early treatment for scurvy. The wood from this exceptionally useful tree was also used by Native Americans to make the frames of canoes which were covered with the bark of white birch, sewn together with tamarack roots and sealed with pine or balsam fir resin.‡

Northern white cedar has a native range that covers the southern part of eastern Canada and adjacent areas of the United States. It can be found in scattered sites as far south as the Appalachian Mountains in Pennsylvania, North Carolina, and Tennessee. It is most commonly found on moist, nutrient-rich sites and is characteristic of seepage areas§ like those found at the bottom of the cliffs and loose talus on our Ontario site. Northern white cedar is long lived, especially on undisturbed sites that retain moisture. It grows slowly, produces the largest quantities of seed only after 75 years of age, and commonly lives to 400 years or older.

Mortality of northern white cedar seedlings is extremely high in their first few years, during which typical growth is only 8 centimeters (3 inches) in the

---

* Burns and Honkala, *Silvics of North America*, vol. 1, *Conifers*.
† Ibid.
‡ Elbert L. Little, ed. *National Audubon Society Field Guide to North American Trees: Eastern Region* (New York: Alfred A. Knopf, 1998).
§ Burns and Honkala, *Silvics of North America*, vol. 1, *Conifers*.

FIGURE 4.4 (See color insert.)
Mature northern white cedar growing on a rock outcropping.

first three years. Drought is the most common cause of seedling mortality.[*] Drought will likely continue to increase in frequency over much of the boreal region of Canada, threatening the natural regeneration of northern white cedar. Protecting the immediate upslope watersheds that provide constant moisture to white cedar stands and maintaining dense cover to

---

[*] Burns and Honkala, *Silvics of North America*, vol. 1, *Conifers*; Little, *Field Guide to North American Trees: Eastern Region*.

increase resistance to drought will assist in their survival. This is one of many examples of how maintaining and protecting a stand of trees or any intact ecosystem is enhanced by familiarity with and knowledge of the specific site and the characteristics of individual species. This is an essential component of what Wes Jackson, of the Land Institute (Salina, Kansas), has characterized as "consulting the genius of the place."[*]

All of the species mentioned here and thousands more have an important role in healthy boreal ecosystems. The ongoing objectives for the site discussed here are to continue to learn, with the assistance of the Ontario Ministry of Natural Resources and local residents, how to restore the approximate mix of primary tree species that existed prior to human disturbance while preserving the relatively intact, healthy ecosystems; selectively harvest valuable species and market them so that they continue to sequester carbon in the form of durable and useful wood products long past the life span of the trees; and appreciate and share the beauty of the place. The time frames involved in pursuing these objectives are certainly longer than my remaining lifetime. I hope that my grandchildren will have an interest in continuing the work that has been initiated, but I do not plan to leave the future preservation of the place to chance. Much of this land will be legally protected from "development."

4.1.2 Northeast Kansas Grassland and Homestead

My family farm in northeast Kansas is also located on a glaciated site. It is about 10 miles north of the Kansas River valley where a pre-Illinoian glacier reached its terminal extent about 700,000 years BP. This area was predominately tallgrass prairie as recently as about 200 years ago. A few remaining undisturbed hilltops have not yet completed the transition to newer deciduous forests, have not been converted to cropland, and remain in native tallgrass prairie.[†] Our 32.4 hectare (80 acre) farm is currently a mixture of approximately 9.3 hectares (23 acres) of eastern deciduous forest on the lower elevations and 20.2 hectares (50 acres) of grass, with the remainder in gardens, ponds, apple and pear orchards, and windbreaks of Austrian and ponderosa pine. There are no native pines in Kansas. The only native conifer is eastern red cedar, which is a vector for cedar apple rust, so I attempt to keep it away from the orchards and orchard windbreaks. Without intervention the woodland tends to expand into the grassland, a process that I have allowed and encouraged in several areas over the past twenty-five years.

The land is located on an upland ridge with virtually no surface water running onto it from adjacent land. Several intermittent streams originate on the farm, four of which are captured by three ponds used for raising fish,

[*] Wes Jackson, *Consulting the Genius of the Place: An Ecological Approach to a New Agriculture* (Berkeley, CA: Counterpoint Press, 2010).

[†] Shane A. Lyle, *Glaciers in Kansas* (Lawrence: Kansas Geological Survey, 2009).

FIGURE 4.5 (See color insert.)
Downspout and cistern for water collection.

irrigating gardens and young planted trees, and watering livestock. The upland ridge and multiple intermittent streams led to the name Creekridge Farm. Household water comes from cisterns that capture rainwater from the roof of the house. Rainwater is also collected in tanks from the roofs of a cattle shed and equipment shed and used for livestock and gardens. A 186 square meter (2,000 square foot) roof in an area with annual average rainfall of 99 centimeters (38.5 inches) will collect about 181,680 liters (48,000 gallons) of water per year. My wife and I use about 37,850 liters (10,000 gallons) per year for household needs and the rest is available for gardens and trees.

This water system was developed more than twenty years ago at approximately the same cost as would have been incurred buying a meter and connecting to the local rural water district. We have no monthly water bills, maintenance costs have been less than $100 in twenty years and the electricity used to run the system is minimal. This option was chosen out of a commitment to live on the immediate watershed and to serve as an example. Students and neighbors have examined our system and several have created stand-alone household/farm water systems adapted to their unique locations and needs. Turning on the tap from a municipal system is much "easier" and, in terms of drinking water, possibly safer than developing a stand-alone water system, especially in urban and suburban environments. Safe drinking water, however, is a very small portion of household water use and an infinitesimally small part of overall water use. Where we obtain our water and how we use it warrants careful consideration.

Reducing the energy and chemical consumption of centralized municipal systems by using treated water for only those purposes for which it is necessary, minimizing stormwater runoff, conserving scarce water resources, and adapting to local resource constraints are likely to become increasingly necessary in the future. Municipal water systems are becoming significant competitors for irrigation water throughout most arid and semiarid regions of the Earth. Irrigation of agricultural land accounts for at least 70% of global water use and the total water available for irrigation in key food production areas is in decline as industrial and municipal use is growing.* Thinking about and initiating less consumptive systems is one of the central requirements of the emerging age of resource scarcity. Each locale and ecosystem presents unique opportunities for individual and collective action to conserve energy and water, build soil carbon and fertility, protect intact ecosystems, and enhance species richness and ecological diversity. These opportunities are unlikely to be acted upon unless they are explicitly recognized as essential, necessary values and concrete examples of feasible, effective avenues for action are evident and encouraged by public policy. Our rainwater collection system is one simple example that is common throughout human history and remains common today in the developing world.

The part of the farm that is now grassland was annually plowed and used to grow mostly corn, soybeans, and oats for at least fifty years prior to being planted to domestic, cool season bromegrass in the early 1980s. It was literally "used" for growing row crops. Because of the highly erodible soils and slopes with 4%–8% grades (think of an 8% grade as a hill just about steep enough for snow sledding); much of the soil was severely eroded and depleted of organic matter and carbon. A high priority in the early years after buying the place was to stop the ongoing erosion that had been slowed but not halted by the then recent conversion by the previous owner of the marginal, excessively sloped cropland to perennial domestic bromegrass. Replacing the bromegrass, which has high nitrogen requirements and a shallow root system, with deep-rooted warm season native grasses was the most important action for long-term erosion control. There were also several areas where gullies had formed and were continuing to expand. Instead of expensive and fossil energy intensive mechanical terracing, the erosion was slowed and eventually stopped using a combination of carefully placed rocks, brush and hay. Loose limestone and granite piled near the edge of what had been crop fields proved to be very useful for this project. Loose rocks found in the woods and those unearthed when building the house and burying an electric line are also being used to line the shoreline of a new pond to prevent bank erosion. One persistent gully created by an intermittent stream

---

* Food and Agriculture Organisation, *The State of the World's Land and Water Resources for Food and Agriculture: Summary Report* (Rome: United Nations, 2011).

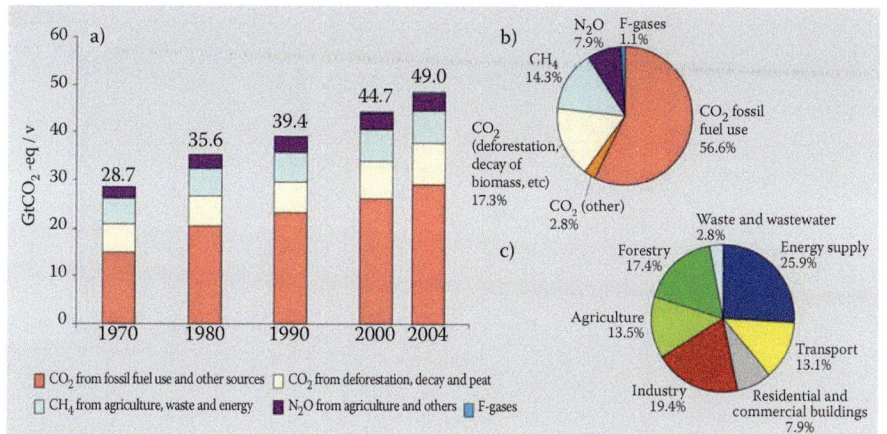

FIGURE 1.1
Greenhouse Gases and Sources. *Source*: IPCC (2007).

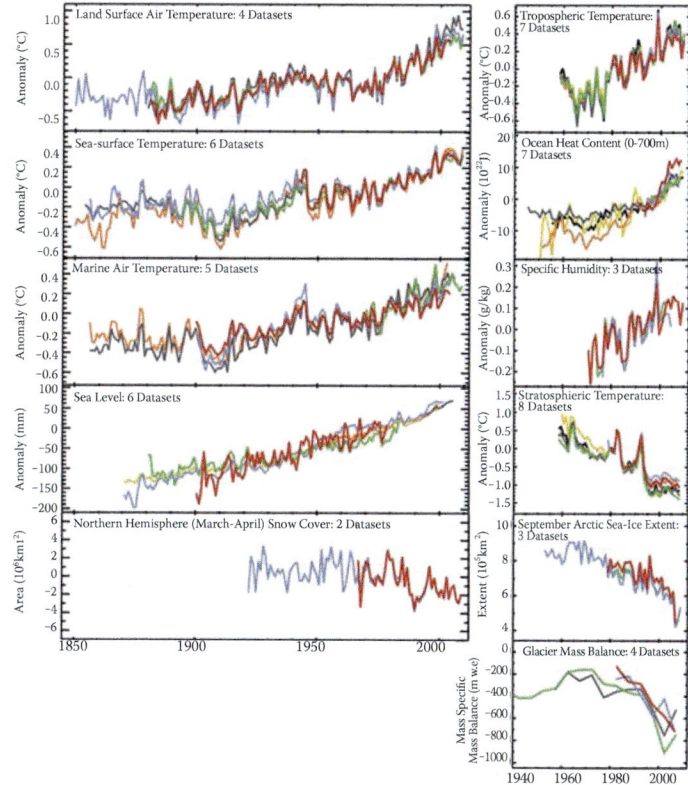

FIGURE 1.2
Warming Indicators. *Source*: National Aeronautics and Space Administration (http://www.ncdc.noaa.gov/bams-state-of-the-climate) D. S. Arndt, M. O. Baringer, and M. R. Johnson, eds., "State of the climate in 2009," *Bulletin of the American Meteorological Society* 91, no. 7 (2009): S1–S224.

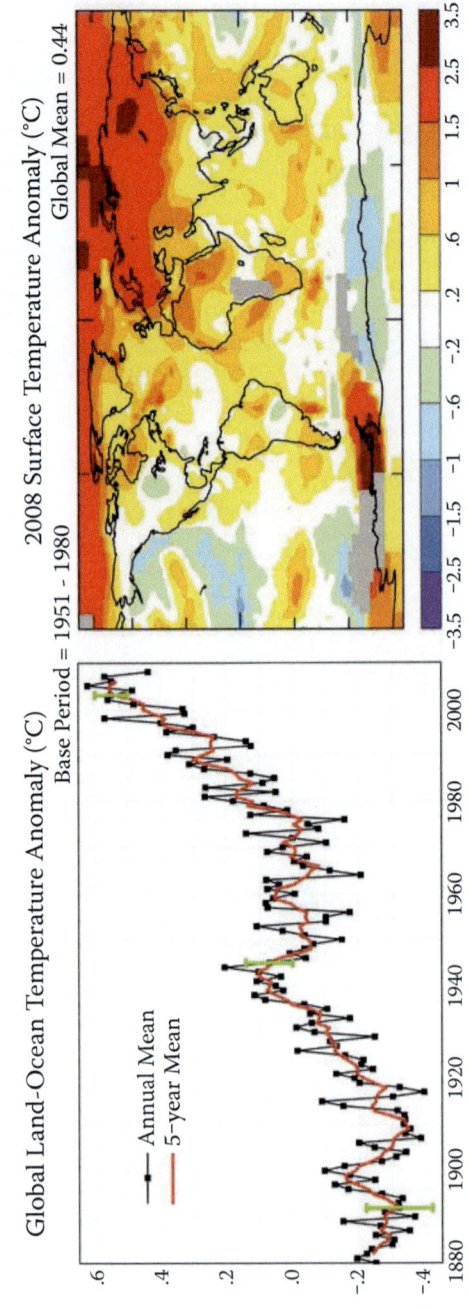

FIGURE 1.3

Global Land–Ocean Temperature Anomaly. *Source:* National Aeronautics and Space Administration, Goddard Institute for Space Studies.

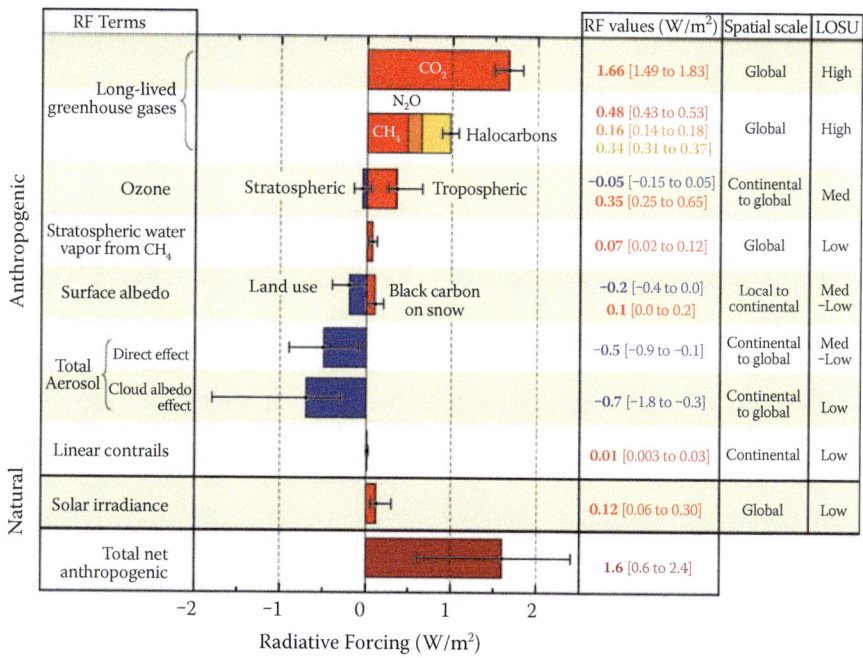

FIGURE 1.5
Radiative Forcing. *Source:* Intergovernmental Panel on Climate Change, 2007.

FIGURE 4.1
Mixed boreal forest with a talus ridge in the background.

FIGURE 4.2
Glacial Lake: "The Hole."

FIGURE 4.3
Young white spruce after a selective harvest of declining aspen and white birch.

FIGURE 4.4)
Mature northern white cedar growing on a rock outcropping.

FIGURE 4.5
Downspout and cistern for water collection.

FIGURE 4.6
Native grasses provide excellent winter soil cover.

FIGURE 4.7
Eastern red cedar pruned as a future timber crop tree.

FIGURE 4.8
Dead branches on live osage orange trees provide seasoned firewood and preserve living trees.

FIGURE 4.9
Horse logging in Michigan. Carl Gillies and his team of Percherons create minimal damage.

FIGURE 4.10
Even a small diesel skidder like this one used in Canada causes significant damage to the forest.

FIGURE 4.11
Refueling the Haflingers.

FIGURE 4.6 (See color insert.)
Native grasses provide excellent winter soil cover.

was repaired with a bulldozer and that stream now feeds a pond that was constructed at the same time.

Slowing the persistent erosion of highly erodible soil on sloping ground requires permanent vegetation and minimal disturbance. In the areas that I planted to native grasses twenty years ago, soil erosion appears to have largely ceased. In the sloped areas that until recently remained a mixture of domestic grasses and encroaching native species, mostly broom sedge and some indiangrass, soil erosion is still evident following heavy rainfall. A recent soil test of those areas measured soil organic matter content at 1.8%. This is primarily Pawnee clay loam with slopes of 4%–8% that typically has an organic matter content near the surface of about 4% in undisturbed ecosystems.* Using the soil organic matter content of undisturbed ecosystems as a standard to work toward, this land must gain substantial carbon before it is fully restored. This area was newly planted to a diverse mixture of twenty-nine perennial grassland species, including native legumes, wildflowers, and forbs, in the spring of 2012.

All of the 20.2 hectares (50 acres) of grass were planted in weak stands of unfertilized domestic grasses, mostly bromegrass, when I purchased the property in 1987. About 12.1 hectares (30 acres) were replanted in 1992 with

---

* U.S. Department of Agriculture, "Map Unit Description, Soils Inventory Report," Natural Resources Conservation Service, Jefferson County, Kansas (2011).

native warm season perennial grassland species on the upland portions of the farm. The warm season native grass planting was done with a no-till drill directly into the existing stand of bromegrass, which had gone unfertilized since it was planted about eight years earlier. The area planted to native perennials was not plowed, disked, or sprayed with chemicals to prepare the seed bed. A local county extension agent informed me that this method would be a waste of expensive seed and would result in poor germination and a thin, weedy stand. I know others who have received and ignored similar advice and now have strong diverse stands of native perennial grassland species. Observations of neighboring farms confirm that following the conventional recommendations regarding seedbed preparation, however, appears to generally produce a strong stand of native grasses more quickly than drilling seed directly into a weak stand of domestic grasses. The trade-offs to consider include fuel consumption, erosion due to tillage on sloped land, and chemical contamination.

The planted mixture of switchgrass, big bluestem, little bluestem, side-oats grama, indiangrass, medium red clover, and Illinois bundleflower was slow to develop for the first three years. After six or seven growing seasons and a well-timed mowing once per year to control weeds, the stand was thick, producing at least twice the biomass previously produced by the bromegrass, and "weeds" (plants growing where you don't want them) were few. The deep-rooted native grassland species mixed with legumes creates a stand that is highly drought resistant and has received no fertilizer for twenty years beyond that distributed by grazing cattle and the nitrogen fixed by the legumes. Long-term plantings of diverse mixtures of native perennial grassland species have been found to be deficient in phosphorus, especially if aboveground biomass is regularly removed for hay or as a bioenergy crop.[*]

A recent soil test of our twenty-year-old stand of native perennials, which as been grazed by cattle in a rotational grazing system with hay harvested every two years, confirmed this finding and showed a slight phosphorous deficiency. Twenty pounds of phosphorus per acre were added to half that acreage for the first time twenty years after planting. The district conservationist with the Natural Resources Conservation Service believes that the phosphorus will enhance overall productivity and help the desirable species compete more successfully against some invading broom sedge, a low-fertility opportunist that is not palatable for cattle. Whether this is true on this site and with the species present will be made clearer over time as areas treated with phosphorous are compared to similar areas left unfertilized. Every field is unique, and an ongoing experimental attitude is helpful when planning

---

[*] David Tilman, Jason Hill, and Clarence Lehman, "Carbon-negative biofuels from low-input high-diversity grassland biomass," *Science* 314, no. 5805 (2006).

grassland or cropland management.* I prefer and recommend avoiding purchased soil amendments as much as possible. The presence of ruminants, in our case low-density cattle grazing, and their manure provides key nutrients. Nitrogen-fixing legumes are a part of native grasslands and should be a component in any pasture system. Avoiding herbicides and insecticides protects microbial health and diversity in the soil (and the health of humans) and helps to ensure the presence of the soil fungi so essential to the exchange of nutrients from soil to fungi to plants.†

Although annual burning of warm season native grasses is common practice in many areas of Kansas, I have found it to be unnecessary for brush control and have burned only small patches on a few occasions. I control invasive eastern red cedar seedlings by cutting them by hand at ground level with loppers. Occasional burning to control cedars and other woody invasives may be more necessary on a larger farm with less labor available per acre. Occasional burning can help control trees and brush. The blackened ground also warms earlier in spring and can give a boost to spring growth of prairie grasses. Annual burning, however, is damaging to nesting bird populations and contributes to air pollution, especially when the smoke mixes with urban ozone pollution. This has become controversial in the nearby Flint Hills region, where many landowners and cattle ranchers traditionally burn native grass pastures every spring.‡

A policy of burning perhaps once every three years to control woody invasive species, minimize air pollution, and have less impact on ground-nesting birds seems reasonable as a compromise and more ecologically sound than annual burning or a total burning ban. Frequent burning of native grass reduces stem and leaf litter and thus likely reduces the accumulation of soil organic carbon. Research on the restoration and management of native grasses is ongoing throughout the world. A central finding thus far has been that grassland ecosystems, even restored grasslands, are exceptionally complex, variable, substantially affected by nonmanagement variables, and will require long-term evaluation to determine effective management strategies.§ Again, as Wes Jackson has wisely counseled, we should "consult the genius of the place" and look to intact native prairies for guidance.¶ As I examine native prairies or look at the experimental results of research comparing the

* Miguel A. Altieri, *Agroecology: The Science of Sustainable Agriculture*, 2nd ed. (Boulder, CO: Westview Press, 1995).
† Janet Raloff, "Environment: chemicals linked to kids' lower IQs: studies identify effects from pesticides still used on farms," *Science News* 179, no. 11 (2011); Adam J. Wargacki et al., "An engineered microbial platform for direct biofuel production from brown macroalgae," *Science* 335, no. 6066 (2012).
‡ Craig Volland, "Flint Hills smoke management plan off to a shaky start," *Planet Kansas* (2011).
§ J. E. Doll et al., "Effects of native grass restoration management on above- and belowground pasture production and forage quality," *Journal of Sustainable Agriculture* 33, no. 5 (2009).
¶ Jackson, *Consulting the Genius of the Place.*

aboveground and belowground productivity of various mixtures of grass-land species,* the feature that stands out is diversity. The more the better.

In my experience the medium red clover that I regularly broadcast into my pastures tends to be crowded out as native perennial grasses become established, while the taller, native Illinois bundleflower that I also plant persists. I am now planting purple prairie clover, another native legume, in both established warm season grass stands and in a new planting in the hope that it will persist with more success than the nonnative red clover and provide more nitrogen fixation and species diversity than the Illinois bundleflower alone. Sometimes it takes a long time to learn a lesson that, in retrospect, should have been apparent earlier.

High-diversity mixtures of perennial herbaceous grassland species are highly productive for pasture or hay crops with little or no added fertilizer. Low-input high-diversity (LIHD) mixtures of native grassland perennials provide more net energy for biofuels, greater net greenhouse gas reductions, and less agrichemical pollution per hectare than corn grain ethanol or soybean biodiesel, even when the grasses are grown on agriculturally degraded land and the corn and soybeans are grown on fertile cropland. Planting a diverse mix of grassland perennials, including legumes, is also a proven way to rehabilitate degraded, nutrient-poor land that has been eroded and depleted of soil organic carbon by being used as cropland.† Even land like the current grassland on our Kansas farm, most of which should never have been plowed and cropped because of the excessive slopes and highly erodible soils, can be restored to productivity and increased carbon storage. This is the central management objective for the Creekridge Farm grassland. The next steps are to continue to expand the diversity by adding more grassland species, expand the acreage planted to native perennials, and utilize the biomass for the production of biofuel.

4.1.3 Biofuels in Ecological Context

A short detour to a discussion of biofuels will help to explain why biofuels production with a diverse mix of perennial grassland species has a place in the future of Creekridge Farm and on other agriculturally degraded lands throughout the world. There are currently two established classes of biomass for energy and biofuels. The first is monoculture crops grown on fertile soil; examples include corn, soybeans, oilseed rape, sugarcane, willow, hybrid poplar, and switchgrass. The second is "waste" biomass; examples include straw, waste wood, corn stover, rice hulls, or other crop residue.‡ The first class of biomass sources has the profound disadvantage of directly competing for scarce fertile land suitable for growing food. If

---

* Tilman et al., "Carbon-negative biofuels from low-input high-diversity grassland biomass."
† Ibid.
‡ Ibid.

we are to expand global food production by the required 70% by 2050 without significantly expanding cropland and thus accelerating deforestation,[*] using significant amounts of scarce fertile cropland for fuels instead of food is simply not feasible.

Waste biomass has more potential than monocrops on fertile soil to make an ecologically sound contribution to a low- or no-carbon fuel supply, depending on the specific biomass used. Removing crop residues from the fields in which they are grown removes organic matter and carbon that is essential to the long-term fertility of the land.[†] Using corn stover, rice hulls, and other crop residues for biofuels depletes soil organic carbon, fertility, and water-holding capacity, and thus also places biofuels in direct competition with food production for the fundamentally scarce resources of soil and water. Occasional removal of a portion of the crop residue from soil with adequate organic carbon may be possible without negative long-term effects. Wood wastes from urban storms, wood processing, and timber stand management is generally in less competition with food production and is a viable source of biomass for biofuels in many parts of the world. Removing too much biomass from a forest or plantation, including the tops remaining after a timber harvest, deadwood, or litter, can also deplete essential soil carbon and contribute to the degradation of forest ecosystems and decrease the long-term productivity of a timber stand.[‡]

Wood from trees damaged by storms in urban areas, invasive and undesirable trees and shrubs removed from farms or timber stands, or used shipping pallets can be and increasingly are being used as biofuel. The supply of these wood wastes often exceeds the demand for mulch, their other primary use, and they commonly end up in landfills. Developing biomass energy and biofuel facilities that utilize wood wastes and minimize transportation distances has the potential to provide economically and ecologically viable use of biomass for energy generation or liquid biofuels. A much greater potential for energy production, positive ecological effects, and food security has been demonstrated, however, for a third emerging class of biofuels, low-input high-diversity (LIHD) mixtures of perennial grassland species grown on degraded agricultural land.[§]

Switchgrass (*Panicum virgatum*) has been identified as a productive perennial bioenergy crop. Switchgrass grown in monocultures on highly productive soil with fertilizers and pesticides is only slightly more ecologically sound, however, than producing biofuels from corn or soybeans

---

[*] Food and Agriculture Organisation, *The State of the World's Land and Water Resources.*

[†] Jackson, *Consulting the Genius of the Place.*

[‡] Chris Maser, *Sustainable Forestry: Philosophy, Science, and Economics* (Boca Raton, FL: St. Lucie Press, 1994).

[§] Christopher B. Field, J. Elliott Campbell, and David B. Lobell, "Biomass energy: the scale of the potential resource," *Trends in Ecology and Evolution* 23, no. 2 (2008); Tilman et al., "Carbon-negative biofuels from low-input high-diversity grassland biomass."

on highly productive soils using fertilizers and pesticides. The same limitation applies to Asian grass hybrids like *Miscanthus* × *giganteum* or several tree species grown in monoculture.* Like corn and soybeans used for biofuels, switchgrass or Asian hybrids grown in this way compete with food crops for increasingly scarce fertile soil. I consider it an indication that the industrial mind has become dysfunctional when even biomass energy crops, intended to replace a portion of fossil fuels, have been largely undertaken using the same industrial model that created the ecological scarcities and degradation that we now face. As we have seen, there is simply not a surplus of soil resources on a global scale to meet the increasing demands for food and, at the same time, grow biomass for biofuels on a significant portion of the Earth's best soils, even if it is a perennial grass. Switchgrass is a perennial, so it does not require annual soil disturbance, and it has lower water and supplemental nutrient requirements than corn or soybeans. Switchgrass in a monoculture sequesters more carbon in its roots and in the soil than corn or soybeans. A much better use of switchgrass, however, is to use it not in monocultures on fertile land, but rather as one component of a diverse mixture of native perennial grassland species grown on agriculturally degraded land.[†]

LIHD mixtures of up to sixteen grassland species, including legumes, are much more productive in terms of biomass production, carbon sequestration, and ecological diversity than the same species grown in a monoculture or in combinations of fewer species. Tilman et al. planted 152 experimental plots of various combinations of 1, 2, 4, 6, 8, or 16 herbaceous perennial grassland species on degraded agricultural land in Minnesota. The plots were not fertilized and were watered only during establishment. The bioenergy production and ecosystem carbon sequestration were measured each year. During the last three years of the ten year experiment the two, four, eight, and sixteen species plots produced 84%, 100%, 157%, and 238% more bioenergy, respectively, than the plots planted in a single species. The LIHD plots with sixteen species also increased their annual biomass production over time more than the monoculture plots or the lower diversity plots. Ecosystem carbon sequestration was higher in the LIHD plots, measured as both soil carbon and belowground root mass, than in the lower diversity plots. Soil carbon storage occurred despite annual burning of the aboveground biomass (done to make accurate annual measurements of biomass production). The most diverse LIHD plots sequestered 4.4 metric tons of carbon per hectare (1.78 metric tons per acre) per year in soil and roots, while monoculture plots were essentially carbon neutral. The researchers estimate that about 5% more carbon may be stored in soils deeper than that measured for the experiment (60 centimeters [23.4 inches]).[‡] This is potentially significant for long-term carbon

---

[*] Tilman et al., "Carbon-negative biofuels from low-input high-diversity grassland biomass."
[†] Ibid.
[‡] Ibid.

storage, because the deeper carbon is buried the more persistent it generally is—centuries to millennia if not disturbed.[*]

The Tilman et al. research demonstrates that most of the carbon sequestered by roots in LIHD perennial grassland plantings occurs in the first decade after establishment. The second decade is estimated to produce about 22% as much root mass as is produced in the first decade. When added to the ongoing soil carbon sequestration, root mass and soil carbon are projected to accumulate at 3.3 metric tons per hectare (1.34 tons per acre) per year during the second decade after establishment of a LIHD mixture of perennial native grassland species.[†]

Planting LIHD mixtures of native grassland species on degraded land meets the criteria identified in Chapters 2 and 3 of a land use or management practice that is inexpensive, low risk, reversible, available immediately, and has multiple benefits. LIHD plantings remove CO_2 from the atmosphere, build soil carbon and fertility in agriculturally degraded soils, utilize land that is unsuitable for cropland, increase resistance and resilience to flood and drought, produce fuel that can replace some fossil fuels, and provide a diverse ecosystem for a wide range of plants, animals, and fungi. Since the plantings are easily reversible and build soil carbon and fertility, it may be possible in ten–fifty years to return some of the restored land back to food production—a way of building and preserving soil for the future and enhancing food security.

Biofuels can be carbon neutral, carbon negative, or net carbon sources when subjected to comprehensive life cycle analysis. In order to be carbon negative a biofuel must result in a net reduction of greenhouse gases. Corn ethanol and soy biodiesel, for example, are both carbon sources—they have net CO_2 emissions. Tilman et al. note that they have lower emissions by 12% and 41%, respectively, than the gasoline and diesel fuel they displace, but remain sources of carbon emissions. LIHD biofuels, in contrast, have been shown to be carbon negative, with net reductions of carbon emissions across the full life cycle of planting, harvesting, transport, and processing. LIHD biofuels are estimated to reduce carbon emissions by 6–16 times as much as corn grain ethanol and soybean biodiesel when used to replace gasoline and diesel fuel.[‡] The net carbon sequestration of LIHD biofuels will vary depending on the productivity of a specific site, the fossil fuels used in production, the distance the biomass feedstock must be transported for processing, the processing method used, and the transport distance of the energy or fuel product.

Tilman et al. estimate that LIHD biofuels grown on available degraded lands could displace about 13% of global petroleum consumption for

---

[*] Brian J. McPherson and E. T. Sundquist, *Carbon Sequestration and Its Role in the Global Carbon Cycle* (Washington, DC: American Geophysical Union, 2009).
[†] Tilman et al., "Carbon-negative biofuels from low-input high-diversity grassland biomass."
[‡] Ibid.

transportation and 19% of fossil fuel used in global electricity consumption. This would eliminate approximately 15% of global greenhouse gas emissions, without accounting for the carbon sequestration in the roots and soils of LIHD plantings.[*] The net reduction of atmospheric carbon would be greater if the experimental results for soil and root sequestration reported by Tilman et al. can be approximated in actual practice in the field. The late-successional native plant species utilized for productive LIHD plantings will vary in each locale around the globe. Widespread implementation will require specific knowledge of local native species, climate, and ecosystems. This is not specialized or rare knowledge.

Extrapolating experimental results, even when an experiment is well designed, is uncertain at best.[†] As noted in earlier discussions of the potential global magnitude of enhanced terrestrial carbon sequestration, actual results depend on the rate and extent to which proven methods are put into practice. LIHD native perennial grassland plantings provide a means of sequestering carbon, rehabilitating degraded agricultural soils, reducing fertilizer and pesticide use, and enhancing wildlife habitat and ecological diversity even if the biomass produced is not used as a feedstock for biofuels. Even if Tilman et al.'s projections of the biomass energy potential of LIHD plantings are overly optimistic, there is overwhelming evidence that biofuels from biomass grown on degraded agricultural lands planted with perennials has great advantages over food crop–based biofuels in terms of greenhouse gases and overall ecological impact.[‡]

A wide-ranging assessment of biomass energy and biofuels in terms of ecological and food security effects, economic viability, and potential impact on fossil fuel use was published by Field et al. in 2008. A central finding was that "the area with the greatest potential for yielding biomass energy that reduces net warming and avoids competition with food production is land that was previously used for agriculture or pasture but that has been abandoned and not converted to forest or urban areas."[§] Biomass was the world's primary energy source prior to the Industrial Revolution and remains roughly equal to nuclear and hydroelectric power as one of the three largest sources of nonfossil fuel energy, at about 7% each, of the total world energy consumption. Wind and solar, by comparison, equaled less than 1% of global energy demand in 2007. Wind, solar, and geothermal energy development has grown rapidly in recent years,[¶] but remains far smaller than biomass globally. Biomass is used for cooking and heating in

---

[*] Tilman et al., "Carbon-negative biofuels from low-input high-diversity grassland biomass."
[†] Michael P. Russelle et al., "Comment on 'Carbon-negative biofuels from low-input high-diversity grassland biomass,'" *Science* 316, no. 5831 (2007).
[‡] Joseph Fargione et al., "Land clearing and the biofuel carbon debt," *Science* 319, no. 5867 (2008).
[§] Field et al., "Biomass energy: the scale of the potential resource," 65.
[¶] Lester Russell Brown, *World on the Edge: How to Prevent Environmental and Economic Collapse,* 1st ed. (New York: W. W. Norton, 2011).

much of the developing world and as a source of industrial heat, mostly in the paper and forestry industries.

The evidence is clear that terrestrial biofuels cannot be considered a replacement for fossil fuels that would allow energy consumption to continue at close to the rate of recent fossil fuel use. Field et al. estimate that if 100% of global harvests of corn, sugarcane, soy, and palm oil were converted to liquid fuels the net total would equal about 1.2% of global energy from fossil fuels. If biomass was grown on all available degraded and abandoned agricultural land, estimated at about 386 million hectares (954 million acres), the total aboveground biomass would represent about 5% of global energy demand.[*]

The greatest value of this form of land use is arguably not the bioenergy produced, but rather the biosequestration of carbon, soil building, and enhancement of species richness and ecological diversity achieved. In an age of climate change and accelerating desertification, grassland species diversity and the multiple ecosystem functions associated with it are especially important in the arid, semiarid, and dry subhumid ecosystems (drylands) that constitute 41% of the terrestrial system. Carbon storage, photosynthetic productivity, and improved nutrient pools have all been found to be positively and significantly related to plant species richness in drylands globally and serve as a buffer against the negative effects of climate change and desertification.[†] The protection, restoration, and creation of diverse mixtures of grassland plant species is an essential component of grassland management and ecological restoration globally.

The primary technical limitation to expanding liquid or gaseous biofuels (cellulosic) from either wood waste or biomass produced with high-diversity mixtures of grassland species is the lack of a commercially viable process to overcome the lignin molecule, needed to effectively extract the sugars to ferment into ethanol. This and other barriers for terrestrial biomass as a source of fuel, including the limited availability of land, water shortages, and the use of fertilizers that emit greenhouse gases, have led to extensive research on macroalgae (seaweed) from marine ecosystems as a feedstock for biofuels. Despite important recent progress in converting seaweed—which contains no lignin and requires no land, freshwater, or fertilizer—to ethanol, significant barriers remain. Alginate, a complex polymer in seaweed, is similar to the lignin in terrestrial biomass in that it is resistant to conversion by industrial microbes. If this barrier is overcome, as it appears it may be,[‡] the fundamental limitations of terrestrial biomass will still apply to marine biomass as well; the sheer volume of seaweed and the long hauling distances required to concentrate the biomass at a processing facility present a hurdle that may

---

[*] Field et al., "Biomass energy: the scale of the potential resource."

[†] Fernando T. Maestre et al., "Plant species richness and ecosystem multifunctionality in global drylands," *Science* 335, no. 6065 (2012).

[‡] Wargacki et al., "An engineered microbial platform for direct biofuel production."

limit ethanol production from seaweed to a volume that will replace only a very small percentage of fossil fuels. It has been estimated that replacing 1% of the U.S. gasoline supply would require growing seaweed on approximately 11,000 square kilometers (4,246 square miles). The cost and energy use to haul seaweed long distances would increase the cost and carbon emissions of the ethanol produced.[*]

Biomass energy used for direct combustion or in emerging liquid and gaseous forms cannot and will not overcome the fundamental physical limitations of mass and space. The energy concentrated in relatively small quantities of fossil fuel will not be fully replaced by diffuse and flowing wind and solar energy sources. The level of energy consumption in contemporary industrial societies is made possible on a temporary basis only by mining the concentrated photosynthetic product produced and compressed over hundreds of millions of years. Biomass energy from the current carbon cycle of photosynthetic production in terrestrial and marine ecosystems will be ecologically and economically viable only in locations where large amounts of biomass are readily available. Biomass energy is well suited to relatively small and dispersed facilities that use locally available materials for feedstock and avoid hauling biomass long distances. To hope or plan for biomass energy to replace more than a small percentage of current fossil fuel consumption is possible only by ignoring available evidence and the finite nature of ecosystems.[†]

Biomass energy from wood waste, high-diversity grassland plantings on agriculturally degraded soils, and marine ecosystems has limited but important potential to provide nonfossil energy in the Anthropocene. The principle emphasized throughout this book—that actions to mitigate and adapt to climate change and ecological degradation are generally sounder if they address more than one dimension of a problem—certainly applies to the use of biomass as an energy source. Biomass energy from high-diversity grassland plantings, wood wastes, and algae will only be viable as a way to replace a portion of fossil fuels and partially meet the need for liquid transportation fuel. Research and development to overcome the lignin and alginate barriers to commercial development continues.[‡] Eliminating liquid fuels entirely from the transportation sector is highly unlikely over the next several decades, so biofuels will likely have a moderate but essential place in the energy systems of the future.[§]

---

[*] Erik Stokstad, "Engineered superbugs boost hopes of turning seaweed into fuel," *Science* 335, no. 6066 (2012).

[†] Field et al., "Biomass energy: the scale of the potential resource," Christopher B. Field and M. R. Raupach, eds., *The Global Carbon Cycle: Integrating Humans, Climate, and the Natural World* (Washington, DC: Island Press, 2004); Tilman et al., "Carbon-negative biofuels from low-input high-diversity grassland biomass."

[‡] Stokstad, "Engineered superbugs boost hopes of turning seaweed into fuel," Erik Stokstad, "Can biotech and organic farmers get along?," *Science* 332, no. 6026 (2011).

[§] James H. Williams et al., "The technology path to deep greenhouse gas emissions cuts by 2050: the pivotal role of electricity," *Science* 335, no. 6064 (2012).

4.1.4 Displacing Coal with Biomass Energy from High-Diversity Grassland Species

Despite the recent popular focus on ethanol and biodiesel, there may be greater immediate opportunities for biomass as a direct combustion fuel. Power plants designed to burn coal can replace up to 10% of the coal they burn with biomass. Using compressed biomass pellets for this purpose may currently be "the most efficient commercial use of biomass energy."[*] In 2011 we enrolled 10.5 hectares (26 acres) of Creekridge Farm grassland in the U.S. Department of Agriculture's (USDA) Biomass Crop Assistance Program (BCAP). The biomass produced will be shipped to a producer-owned, nonprofit biomass cooperative in Centerview, Missouri. Biomass from our high-diversity perennial mixture of grasses and legumes will be converted into compressed cellulosic pellets and marketed to generate electricity and displace coal in regional commercial power plants. The pellets can also be used for heating, cooking, or a variety of other direct combustion purposes. The development of this biomass processing plant and others like it may be a precursor to producing cellulosic liquid or gaseous fuels as that technology develops. Our processor is working to develop that capacity.

The BCAP was included in the 2008 U.S. Farm Bill to provide incentives to farmers, ranchers, and forest land owners to establish and produce biomass crops. A thirty-nine-county area in central and western Missouri and eastern Kansas, which includes our farm, was the first BCAP area to be established in the United States. Enrollment of eligible land began in 2011 and is projected to eventually include up to 20,235 hectares (50,000 acres) of land. More than 8,094 hectares (20,000 acres) were enrolled in the first year. The program pays an annual per-acre rent for a period of five years, 75% of biomass crop establishment costs (including preparation and planting costs, seed, soil tests, and fertilizer, if needed), and a payment matching the price received for the first two biomass crops sold during the five-year contract. The annual rent payments are $75.43 per acre per year for our low- to marginal-quality soil types and the total establishment costs are estimated at approximately $200 per acre. In our case, 5 hectares (12.5 acres) of the mixed grassland species that I had planted twenty years earlier qualified for the program with no planting or weed control required. An additional 5.5 hectares (13.5 acres) were planted in the spring of 2012 with the Natural Resources Conservation Service's recommended BCAP mixture of three native warm season grasses (big bluestem, indiangrass, and switchgrass) and two native legumes (Illinois bundleflower and purple prairie clover). The district conservationist agreed that I could, at my own expense, add additional species of grasses and prairie flowers to increase the species diversity of the planting. I added four grass species

---

[*] Field et al., "Biomass energy: the scale of the potential resource," 67.

and twenty native flowers and forbs for a total high-diversity planting of twenty-nine species.

We plan to sell our first two biomass harvests in 2015 and 2016. The energy plant is a distance of 169 kilometers (105 miles) from our farm. This is a longer-than-ideal distance to haul an estimated 60 metric tons (132,000 pounds) of hay. If facilities like our processing cooperative prove successful and additional plants open, travel distances for biomass energy in our area will decrease. Careful records are being kept of all fossil energy use for preparation, planting, and mowing for weed control and will be kept for harvesting, transportation to the processing plant, and transportation from the plant to the final destination of the pellets.

The 60 metric ton estimate for gross production on 10.5 hectares (26 acres) is based on an estimate of 2,269 kilograms (5,000 pounds) of baled weight per acre, consistent with the yield from native grasses on the farm in the past. Selling the biomass by weight, with moisture content measured upon delivery, will correct any errors of estimation. Harvested late in the year, the moisture content should be approximately 10% by weight, which means a total delivered dry weight of about 54 metric tons (119,000 pounds) from the 10.5 hectares (26 acres) in biomass production.

How much coal will 54 metric tons (119,000 pounds) of dry biomass from 10.5 hectares (26 acres) of degraded farmland displace? The bituminous coal burned in U.S. coal plants averages about 78% carbon.[*] The carbon content of biomass is surprisingly constant across a wide variety of plant types and species at about 45%–50% by dry weight. A commonly used average for biomass is 47.5% carbon.[†] One pound (0.45 kg) of carbon from coal or biomass is comparable and produces an approximately equal amount of heat when burned. At 47.5% carbon, 54 metric tons (119,000 pounds) of dry biomass contains about 25.65 metric tons (56,525 pounds) of carbon. This is equivalent to 32.88 metric tons (72,476 pounds) of coal, assuming the coal has 78% carbon content.

If these estimates prove accurate, each year of production of biomass energy with a diverse mixture of perennial grassland species on just 10.5 hectares (26 acres) of degraded agricultural land can displace about 33 metric tons of coal and keep 25.65 metric tons of fossil carbon (94 metric tons of CO_2) from entering the atmosphere. In addition, if the carbon sequestered by the roots and soil of these 10.5 hectares equals the estimate of 1.78 metric tons per acre measured in the experiment by Tilman et al.,[‡] an additional 46.28 tons of carbon (169.84 metric tons of CO_2) will be removed from the atmosphere each year. These rates of carbon sequestration will continue for at least a decade and the removal of carbon from the atmosphere will continue at a lesser rate in the subsequent decade and beyond. When the

---

[*] B. D. Hong and E. R. Slatick, "Carbon dioxide factors for coal," *Quarterly Coal Report* January–April (1994), http://www.eia.gov/cneaf/coal/quarterly/co2_article/co2.html.

[†] Food and Agriculture Organisation, "Carbon content estimation," http://www.fao.org/forestry/17111/en

[‡] Tilman et al., "Carbon-negative biofuels from low-input high-diversity grassland biomass."

carbon sequestered annually and the fossil carbons displaced annually are summed, the total annual displacement of fossil carbon from the atmosphere is approximately 72 metric tons (264 metric tons of CO_2). This is comparable to the amount of carbon or CO_2 emitted from burning about 28,852 gallons of gasoline, or what would be consumed by eighty-seven automobiles traveling 10,000 miles at 30 miles per gallon. (See Appendix A for conversion units.)

As a reminder, CO_2 from the combustion of biomass is from the current terrestrial photosynthetic carbon budget and does not add significantly to atmospheric CO_2 beyond that which would be returned to the atmosphere naturally in a few years if the biomass were left to decay. The CO_2 emitted from burning coal, on the other hand, is from carbon sequestered in the geological sink for hundreds of millions of years and its release into the atmosphere represents an interruption of the natural terrestrial and marine carbon cycle.

The example of carbon sequestration by perennial grassland species and the fossil carbon displaced by biomass fuels from grassland cellulose also serves to illustrate the comparative magnitude of carbon sequestration in forest and grassland ecosystems. One hectare of established forest stores an average of 155, 239, and 242 metric tons of carbon per hectare in temperate, boreal, and tropical biomes, respectively.[*] Preserving the large stores of carbon currently in forest biomes, which take decades to centuries to establish, logically take priority in terms of carbon storage. Most agricultural soils have lost from 30% to 75% of their antecedent soil organic carbon pool, or about 30–40 metric tons per hectare (12–15 tons per acre).[†] Active management to restore lost soil organic carbon and productivity to degraded agricultural land by restoring diverse mixes of native perennials on as much land as feasible will help reduce atmospheric carbon pollution. Equally important, restoration of degraded agricultural land builds soil for future food production and enhances species richness and ecological diversity. Forest and wetland ecosystem preservation and restoration in combination with the restoration of carbon in degraded agricultural soils constitute the core of sustainable land management in the Anthropocene epoch.

4.1.5 Kansas Woodland

Contrary to popular stereotypes about flat Kansas cropland, much of the eastern third of the state is eastern deciduous forest and hilly—rolling prairie mixed with forest is an apt description. Kansas has 850,000 kilome-

---

[*] Pan et al., "A large and persistent carbon sink in the world's forests."
[†] R. Lal et al., "Soil carbon sequestration to mitigate climate change and advance food security," *Soil Science* 172, no. 12 (2007).

ters (2.1 million acres) of forestland.* The primary trees in the 9.3 hectares (23 acres) of woodlands on Creekridge Farm are bur oak, chinkapin oak, northern red oak, shagbark hickory, honey locust, hackberry, American elm, red elm, redbud, osage orange (hedge), red mulberry, black walnut, green ash, cottonwood, and eastern red cedar. There are also a few American basswoods and bitternut hickories and many vines, shrubs, and herbaceous understory plants. The glaciers had a defining impact on the landscape, soils, and vegetation in this and all glaciated regions, with great local variability. The glacier terminating in Kansas deposited erratic pink and gray granite rocks and boulders, known locally as Wisconsin granite, in the northeastern Kansas glaciated hills. They are identical in appearance to the granite found in the Canadian forest west of Lake Superior and create a unique and intriguing geological connection between two sites 1,452 kilometers (900 miles) apart.

Much of the Creekridge woodland is on Vinland complex soils that have a root restrictive layer of bedrock at a depth of 26–51 centimeters (10–20 inches). These areas also have slopes of up to 15%. Despite not being ideal for high-quality timber production, these woods have a wide diversity of species and there are enough openings in the bedrock that large, long-lived bur, northern red, and chinkapin oaks are growing in several stands. There are also a few areas, mostly along streambeds, with deep soil that support valuable black walnut. The primary timber crop trees are black walnut, bur oak, and northern red oak. The areas containing these species are managed by pruning young potential crop trees and removing less desirable species that are in competition for crown space.

These practices are part of a forest stewardship plan developed in cooperation with the Kansas Forest Service, which provides the technical assistance of a professional forester in the development of a long-term forest management plan. Each U.S. state has similar resources available to landowners interested in preserving and improving their woodlands. A 3.48 hectare (8.6 acre) portion of the woodland is also enrolled in the Environmental Quality Incentives Program (EQIP) for Forest Health. This program pays a small amount, about $729 per hectare ($295 per acre) in our case, for intensive timber stand improvement recommended and supervised by a district forester to improve forest health. This program is also available throughout the United States and is administered by the Farm Service Agency and the Natural Resources Conservation Service. Our plan includes the management practices of pruning valuable oak and walnut species, removing trees that compete for space with future crop trees, and removing vines that impede tree growth.

---

* David N. Bruckerhoff, "Improving Your Woodland for Timber Production," L-725, Kansas State University Agricultural Experiment Station and Cooperative Extension Service, Manhattan, Kansas (2007).

FIGURE 4.7 (See color insert.)
Eastern red cedar pruned as a future timber crop tree.

We are also trying something unusual in this area by very lightly thinning a couple stands of eastern red cedar (*Juniperus virginiana*) and pruning their branches to a height of about 3 meters (9 feet) to promote the development of sawlogs. The thinning involves removing only those trees with damaged or forked trunks. These conifers grow well and produce clear stems with a closed canopy (their crowns touch neighboring trees).[*] Eastern red cedar has been widely used as a windbreak tree and is viewed by many as an invasive weed in pastures and woodlands. Its wood is highly valued by some woodworkers for its beauty, durability, and workability. This may present an opportunity for many landowners, as eastern red cedar is the most widely distributed conifer of tree size in the eastern half of the United States. It is very common on abandoned farmland and as a pioneer species in mined areas. Eastern red cedar is drought resistant, adapts to dry environments, can withstand extremes of heat and cold, and has a fibrous root system that is effective at holding shallow and degraded soils in place. It is fast growing and commonly reaches a height of 12–15 meters (40–50 feet) and a diameter of 30–61 centimeters (12 to 24 inches) in about fifty years. The fruits are high in fat, fiber, and carbohydrates and provide food for a range of birds and other wildlife.[†] Eastern red cedar, which I had once considered a nuisance

---

[*] Burns and Honkala, *Silvics of North America*, vol. 1, *Conifers*.
[†] Ibid.

as a pasture invasive, may prove to be increasingly useful ecologically and commercially in an era of abandoned farmland, depleted soils, and more frequent drought.

Black walnut (*Juglans nigra*) is one of the most valuable and scarcest of North American hardwoods. The wood is used for fine furniture, paneling, and a variety of specialty items, including gunstocks. The supply of high-quality black walnut has been diminishing since at least the 1980s and is now used primarily for veneer. Because of its high value and scarcity, black walnut is often the focus of intensive management in eastern deciduous forests. It is intolerant of shade and responds well to pruning and crown release. Black walnut trees are also the focus of much of the pruning and thinning done as part of our timber stand improvement work. The growth rate and overall health of black walnut is improved by removing competing trees so that the crowns of nearby trees are at least 1.5 meters (5 feet) from the crown of the future crop tree. Pruning small lateral branches when the tree is a sapling and carefully removing dead branches without damaging the branch collar will help to produce knot-free wood and protect the health of the tree from damaging rot.* The management strategy for bur and other oaks, which are also intolerant of shade, is similar.

Thinning and pruning to encourage the development of high-quality timber in the future is not generally motivated by individual monetary gain. A black walnut or bur oak sapling pruned in 2012 may be of harvestable size by 2082 or after. Work done now could have monetary benefit to someone in the future, and this may be sufficient as a reason for some to do the work. One of the long-standing conflicts between those who profit from timber harvests and environmentalists and ecologists concerned with the health of forest ecosystems has been the "rotation time," the time between harvests. This is especially pronounced in the Pacific Northwest where the life span of Douglas fir, for example, is commonly 400–500 years and may be 800 years or more. An 800-year-old tree can take 400 years to decompose—a 1,200 year cycle that is equal to sixteen human lifetimes of seventy-five years. By managing forests to shorten the lives of trees, humans have altered long-term ecological processes in ways that are little understood. "To maximize the harvest of wood fiber today, we gamble with the existence of the forests of tomorrow."†

Replacing most clear-cutting with selective harvesting and allowing some trees to live their entire life span and to decay in place will increase the security and duration of carbon in forests and help to maintain ecological diversity. Information about the reproduction and early growth characteristics, reaction to competition, life span, and a range of other characteristics of individual tree species is readily available (see the citations in this chapter). The

---

* Burns and Honkala, *Silvics of North America*, vol. 2, *Hardwoods*.
† Maser, *Sustainable Forestry: Philosophy, Science, and Economics*, 71.

care and management of woodlands, grasslands, croplands, or a residential lots can be enhanced by recognizing and enacting a fundamental ecological rule of thumb: plant or maintain a diversity of long-lived perennial species that are native to or well adapted to a place. Enhanced carbon biosequestration, strong drought and flood resistance, and an improved ability to avoid chemical contamination will generally follow.

The long time frames involved in sound forest management, even of species that mature to harvestable size in 50–100 years, works against an orientation toward maximizing profit as quickly as possible. A couple other reasons to do this work, however, seem more than sufficient. A forest that produces high-quality timber is more likely to have a portion of its biomass preserved in durable wood products long after the life span of the trees than in unmanaged or overexploited woodland with low-quality timber. A forest that produces valuable timber products is also more likely to avoid deforestation and net carbon emissions than a forest used only for fuelwood. Globally the lowest rates of deforestation occur in regions with high rates of timber harvest and forest product output.[*] When forests are of low quality due to overexploitation or neglect the value of alternative uses is often perceived to be greater than the value of preserving the forest.

More important than pruning and thinning to encourage the development of high-quality timber in eastern deciduous forests is management undertaken to maintain the overall health and diversity of the forest. The idea that many forests are healthier when appropriately managed than when "left alone" is counterintuitive to many. In considering this issue it is important to make a distinction between pristine old growth forest ecosystems that have had little direct disturbance from humans and the vast majority of global forest land that has been harvested repeatedly, fragmented, and invaded by exotic species introduced by human settlements. The former are rare and most valuable as ecosystem preserves protected from disturbance. Most of the rest have been fundamentally altered by human actions and will benefit from careful restoration and management. In our small eastern deciduous woodland, that means maintaining species diversity, controlling invasive species, planting trees in areas of weak regeneration, and improving the quality of future timber crop trees.

Our house is heated with wood using a cast iron stove equipped with a catalytic converter. Firewood is harvested primarily from dead lower branches of osage orange (*Maclura pomifera*) and from dead trees, mostly elm, a majority of which tend to die in our area when they reach approximately 20 centimeters (8 inches) in diameter. Occasionally an oak, hickory, or locust deadfall, or a tree removed for timber stand improvement will be used as well. Osage orange (also known as hedge, hedge apple, bodark, bois-d'arc, bowwood and naranjo chino), like eastern red cedar, is commonly considered an invasive

---

[*] Kansas Forest Service, "Loss of forests: development or timber harvest," *Kansas Canopy* Winter, no. 41 (2011).

FIGURE 4.8 (See color insert.)
Dead branches on live osage orange trees provide seasoned firewood and preserve living trees.

species in pastures and woodlands. It is also considered by many to be the best, longest-lasting firewood available. Knowledge of some of the uses for this versatile species can lead to an appreciation of its unique characteristics and value, especially as warming increases and drought becomes more common. The natural range of osage orange is the Red River drainage of Oklahoma, Texas, and Arkansas and surrounding areas of northeastern Texas. It has been planted as a hedge in each of the forty-eight coterminous states and in southeastern Canada. It is most prevalent in the south-central U.S. states and Great Plains region east of the Rocky Mountains. This tree has been credited by some as making agricultural settlement of the prairies possible. It was used as a living fence before the invention of barbed wire and then provided the wood for most of the posts for the wire fences that followed.[*]

Osage orange heartwood, bark, and roots contain many extractives of potential value in food processing, pesticide manufacturing, and dye making. It makes a sturdy landscape tree resistant to wind and ice damage, road salt, heat, and urban air pollution. It has abundant powers of natural regeneration on nearly any soil type and quickly invades exposed eroding soil, especially in overgrazed pastures. It will grow on soil too alkaline for most forest trees. Osage orange tolerates extreme drought even though its native range

---

[*] Burns and Honkala, *Silvics of North America*, vol. 2, *Hardwoods*.

was primarily in moist soils, where it thrives. It also grows and regenerates in shade, although it thrives in direct sunlight. It is one of the healthiest tree species in North America, and in trials it has sustained less damage from insects, diseases, drought, hail, and ice, and, along with bur oak, survived better on uplands than any other deciduous species in the southern Plains.*

Osage orange heartwood is the most decay resistant timber in North America and is immune to termites. About 3 million osage orange fence posts were sold per year in Kansas during the 1970s.† There are still solid hedge fence posts on our Kansas farm that, according to the previous owner, have been in place for more than seventy years without rotting. Posts that I set twenty-five years ago appear nearly new. A person with access to an osage orange stand willing to do hard work and fight their stiff thorns can sell as many posts as he or she can cut for about $8–$10 per post in 2012. The stumps of young trees of post size (10–26 centimeters [4–10 inches] in diameter) will quickly regenerate multiple stems that can be pruned (coppiced) to produce another harvest in ten–twenty years, depending on the site. Prairie farmers of an earlier age commonly clear-cut single-row osage orange hedges on a ten–sixteen year cycle, producing about 2,500 posts per kilometer (4,000 per mile).‡ Perhaps if we make the transition to a more ecologically sound society and economy that places a high value on resource productivity and less value on labor productivity (i.e., labor replacement),§ and a high value on durable wood products replacing fossil fuel–intensive steel and plastic, this kind of labor and natural resource use may once again become common.

Creekridge Farm produces grass-fed beef grown without added hormones or antibiotics. About six to eight head are slaughtered each year by a local state-inspected processor, cut, frozen, and marketed through a local grocery store and directly to customers, mostly neighbors, from the farm. The orchards include approximately sixty apple trees and thirty pear trees, about half of which were of bearing age as of 2012. There are multiple varieties of apples and pears, selected for disease resistance and diversity. No synthetic pesticides or fungicides are used on the fruit trees. Moth larvae, a common pest on apples east of the Rocky Mountains, are controlled by spraying *Bacillus thuringensis* (BT), a naturally occurring bacteria. Organic soybean oil is used as dormant oil to control scale, mites, and a few other pests. The apples and pears are used to make sauce, juice, and for fresh eating by our extended family. Any surplus is sold to a local grocery store or through a local producer's cooperative.

We also grow cherries, peaches, blueberries, hazelnuts, quince, currants, and currant–gooseberry crosses for family use. Some of these perennial crops

---

* Burns and Honkala, *Silvics of North America*, vol. 2, *Hardwoods*.
† Ibid.
‡ Ibid.
§ Tim Jackson, *Prosperity Without Growth: Economics for a Finite Planet* (London: Earthscan, 2011).

may be expanded for marketing in the future. From 1993 through 2003 we grew and marketed annual vegetables and cut flowers through the Rolling Prairie Farmers Alliance, a local producer's cooperative/food subscription service. Starting in 2004 we began the transition from annual vegetable production to perennial food crops and herbs, maintaining a small annual vegetable garden for household use. This greatly reduced the need for fossil-fueled, mechanized tilling and soil disturbance.

Cattle manure composted with grass hay, kitchen wastes, and other plant material from the gardens is used as fertilizer in the orchards and gardens. Legumes, including clovers, peas, and vetch, are also planted in the gardens and around trees for additional nitrogen and to improve tilth. Small amounts of liquid fish emulsion fertilizer and dried kelp are used to provide trace minerals in the orchards and gardens. The dried kelp mixed with salt is also provided to the cattle in place a commercially prepared mineral supplements. Following a recommendation made by Joel Salatin, we use Shaklee Basic H as a cattle wormer (a use not endorsed by the Shaklee Corporation). Basic H is a biodegradable soap made from two soybean enzymes; it is effective and allows us to avoid the chemical insecticides used by feedlots and most conventional beef producers. Basic H provided at a rate of 0.275 liter (½ pint) per 378 liters (100 gallons) of water twice each year, keeping the cattle out of ponds and other riparian areas, providing clean water at all times, and rotating animals on a frequent basis to clean pastures with a diverse range of forages has proven to be an effective system for producing healthy animals.[*]

The management objectives on Creekridge Farm can be briefly summarized. They include continuing to reduce and eliminate erosion on the slopes of the grassland; and building soil carbon and fertility by expanding the extent of and improving diversity in the perennial native grass and legume stands; protecting and increasing the carbon stored in the woodland by maintaining diversity in the forested areas and improving the timber stand by pruning and thinning in areas containing potential timber crop trees; and increasing production and income from food and energy products while reducing fossil fuel use and soil disturbance. Each activity addresses more than one objective. Increasing species richness secures carbon. More biomass and soil organic carbon improves fertility and provides resistance and resilience to drought. Increased biomass on the portion of the grassland devoted to energy production increases belowground carbon storage and income. Improved timber quality increases the value of the woodland for future income potential and for long-term carbon storage in wood products. Each of these activities and objectives also provide a sense of purpose and pleasure in thinking about and working on the land in ways that have a mitigating and adaptive impact on climate change and ecological degradation.

---

[*] Joel Salatin, *Salad Bar Beef*, 1st ed. (Swoope, VA: Polyface, 1995).

4.1.6 Mature Lower Michigan Deciduous Forest

A third piece of land that I own with my brother is located in the southeast-ern portion of Lower Michigan. This land too was profoundly shaped by the Wisconsinan glacier, which withdrew from southern Michigan about 35,000 years BP. Like northwestern Ontario, the Lower Peninsula of Michigan was repeatedly glaciated, but the glacial deposits were not exposed to the repeated torrential bursts of water that affected northwestern Ontario from 9,500 BP to 5,500 BP. This and other differences, including the longer elapsed time since the last glaciation, have left relatively deep soils and few exposed rock outcroppings in the Lower Peninsula, except along some shorelines of Lake Huron in the east and Lake Michigan to the west. Glacial lakes were common throughout Michigan, which now has more than 5,000 lakes and thousands of peat-producing wetlands. Some of the wetlands in Lower Michigan have peat deposits up to 23 meters (75 feet) thick and are estimated to have formed over a period of at least 15,000 years.*

Our 61.1 hectares (151 acres) of Michigan land has approximately 4 hect-ares (10 acres) of peaty wetland that remains wet throughout the year and at least 6 more forested hectares (15 acres) that are waterlogged most seasons and have deep, black, carbon-rich soils. The rest of the acreage is mostly well-drained, deep sandy loam. About 21 hectares (52 acres) of previously cropped ground are now enrolled in the Conservation Reserve Program (CRP) and planted in perennial grasses. About 8.1 hectares (20 acres) of the former crop-land is planted to native warm season varieties and the remainder in a mix-ture of domestic orchard grass and timothy. Nearly all of the area planted to grass is relatively level and erosion is now minimal. There are about 36 hectares (90 acres) of forested land, including several acres of abandoned apple orchards planted in about the 1930s. The primary tree species are red maple, sugar maple, black cherry, white oak, northern red oak, American beech, bigtooth aspen (poplar), bitternut hickory, and cottonwood. There are a few ironwoods, blue beech, and a range of understory plants. This is high-quality eastern deciduous forest with deep, relatively fertile soils compared to the woodlands in eastern Kansas. Although several of the tree species are the same as those found in northeastern Kansas, the size and quality of the timber is far superior.

When we obtained the Michigan farm in 2006 from an uncle, the forest-land had been used as a hunting preserve and had gone almost entirely untouched for at least seventy-five years, probably longer. There was no evi-dence of timber harvesting, pruning, or thinning other than the telltale signs of stump regrowth from long ago. The trees growing from the stumps of previously harvested trees, many with multiple stems, were estimated by experienced foresters to be at least seventy-five years old. The canopy in most

---

* Stanard C. Bergquist, *The Glacial History and Development of Michigan* (East Lansing: State of Michigan, ND).

of the forest was closed by the crowns of mature trees. A closed canopy has complex effects in a mixed-species, uneven-age stand. The predominant and most commercially valuable species in this forest, including black cherry, white oak, northern red oak, sugar (hard) maple, and red (soft) maple, each respond differently to a closed canopy in terms of seedling regeneration, sapling survival, and growth rates throughout their life span. Similarly, each species responds differently to the varying light levels associated with different harvesting and thinning practices. Black cherry, for example, will germinate well in shade and produce a large number of seedlings under a dense canopy, but very few grow to more than 12–15 centimeters (5–6 inches) or survive more than five years with a low level of light. If the canopy is opened only slightly by single-tree selection and harvesting, black cherry reproduction will be poor. A more significant harvest to the point of near clear-cutting will result in extensive regeneration of black cherry. Black cherry is classified as intolerant of shade, as it will not live for extended periods or grow to larger sizes without moderate to heavy opening of the canopy.[*] Black cherry will be gradually eliminated from a mature stand of trees that is not harvested or thinned for long periods.

White oak produces fewer seedlings than most other oaks, but they can persist for many years in the understory by dying back and resprouting. This allows them to grow more rapidly and compete well with other vegetation when the canopy is opened by thinning or harvesting. Once again, white oak will not grow without a relatively open canopy. Opening the canopy in at least parts of the forest that have adequate advance regeneration is a recommended silviculture practice for managing white oak stands. These are large, valuable, trees that can live for up to 600 years.[†] We want to keep them as part of this mixed species forest.

Red maple and sugar maple are much more shade tolerant than the oaks and black cherry. Red maple is also much more tolerant of wet soil and is the dominant species in the woodland swamp. Red maple also has the shortest life span of the valuable species mentioned. All of these factors, and several more, are important in considering the harvesting and management approach to a stand of high-quality eastern deciduous forest. In approaching this task my brother and I had some knowledge and experience gained from managing forest land in Kansas, Ontario, and New York. The mix of species, relatively high-quality site, and length of time since previous harvesting or thinning presented conditions unique to this specific woodland. Deciding to proceed cautiously, we first consulted and walked the woods with a Michigan state forester. The forester provided some general information about the forests in Lower Michigan and a list of timber harvesters and sawmills in the area. We had three timber buyers representing commercial mills mark the trees they proposed to harvest if

---

[*] Burns and Honkala, *Silvics of North America*, vol. 2, *Hardwoods*.
[†] Ibid.

offered a contract and each offered a bid. The monetary bids and the harvest proposal from each were similar. They all proposed to harvest essentially every tree 36 centimeters (14 inches) in diameter at breast height (DBH) or larger, except for a few overly mature individuals that were of little value as timber. They also proposed to complete the harvest over a short period of time using multiple, diesel-fueled timber skidders and acknowledged that there would be substantial collateral damage to young trees in the woods. We decided to accept none of the offers.

Over the next two years we visited with other landowners in the area, several more timber buyers from sawmills, an independent contractor who was given an opportunity (and failed the test) to clean up some timber and firewood removed from a utility right-of-way, and again decided to pass on all offers. In the meantime I had begun reading about silviculture, defined as "the management of forests or woodlands for the benefit of the entire ecosystem, regardless of whether the land is being exploited commercially for the production of timber and other wood products,"[*] and became aware that there are timber harvesters who use horse teams instead of diesel skidders and who do more than pay lip service to the concept of "sustainable forestry." With the assistance of the state forester I had contacted earlier, I located Carl Gillies, a horse logger from Lower Michigan who is passionate about managing forestland for long-term health and diversity. Carl also participates in ongoing continuing education on best management practices in sustainable forestry provided by Michigan State University.

Carl has thirty years of experience harvesting timber and gave up his skidder for horses in about 1990. He usually works alone with his horses, chainsaws, a small tractor, and a device called a clam trailer that is equipped with a small boom for loading logs. The trailer is used for some logs that are far from the landing, when the horses need rest, and for days spent primarily cutting without the horses. Carl has a two-horse team of Percherons and a pair of Haflingers. Percherons are a large breed of draft horse easily capable of dragging, without wheels, logs weighing at least 1,134 kilograms (2,500 pounds). The Haflingers are more compact, but also remarkably powerful and durable animals. The fossil fuel used to transport a horse trailer approximately 32 kilometers (20 miles) to our woodland and Carl's other small equipment is a fraction of the fuel that would be consumed by multiple diesel log-skidding machines that use approximately 76 liters (20 gallons) of fuel each in an 8 hour workday. The logging trails made by Carl, his horse teams, and the small tractor used to occasionally pull the clam trailer are 2 meters (6 feet) wide, much narrower than those made by log skidders. Most importantly, the horses and small equipment are handled to minimize soil disturbance and damage to seedlings, saplings, and pole size trees that will remain when the harvest is complete.

---

[*] Allaby, *A Dictionary of Ecology*, 3rd ed. (Oxford: Oxford University Press, 2005), 398.

FIGURE 4.9 (See color insert.)
Horse logging in Michigan. Carl Gillies and his team of Percherons create minimal damage.

The first thing we learned when we began working with Carl was that many of the smaller trees marked by the timber buyers would stay in the forest. Carl generally refused to cut them, for both ecological and economic reasons. We learned that a well-formed and healthy sugar maple, white oak, or black cherry, for example, that is 36 centimeters (14 inches) in diameter is more valuable aesthetically, economically, and as part of the forest ecosystem if it is left standing than if it is harvested. A white oak that could easily live for more than 300 years and likely grow vigorously to at least 125 years should not be taken, in Carl's view, at 60 years if the forest is truly being managed for long-term health and value. Once this ethic is adopted, an aesthetic appreciation for the forest and individual trees can cause decisions about taking or leaving trees to become difficult. Carl prefers to err on the side of leaving more trees rather than fewer.[*]

Our forest will have more trees and less damage to remaining trees and understory vegetation than would have been the case had we accepted one of the bids from the commercial mills. Some areas of the woods have been cut heavily and the canopy is open enough to allow regeneration of species intolerant of shade, such as black cherry and white oak. Some areas have been harvested more lightly and the canopy remains relatively intact. Time will tell if this varied method will result in successful regeneration of the

---

[*] Carl Gillies, personal communication, February 2, 2012.

FIGURE 4.10 (See color insert.)
Even a small diesel skidder like this one used in Canada causes significant damage to the forest.

FIGURE 4.11 (See color insert.)
Refueling the Haflingers.

several species that have quite different light and space requirements. The complexity of interacting variables in a mixed-age, species-diverse eastern deciduous forest is daunting and not amenable to a simple one-size-fits-all management strategy.*

Other differences in working with Carl Gillies, or someone with a similar approach, involve the interrelated aspects of the time frame for the harvest and the marketing of harvested timber. Each of the mills that provided bids to harvest our timber proposed to complete the job in a period of two–three months. Carl is now in the fourth year of a harvest process that will likely run for a total of six–seven years. This is not primarily because horse logging is so much slower than logging with multiple diesel skidders, although it clearly takes more time. Carl sells logs directly to two small local woodshops that make furniture, cabinets, hardwood flooring, molding, and varied other finished or semifinished wood products. Both of these shops cut and kiln dry their lumber and are willing to pay a premium over commercial prices to obtain the quality and quantity of logs they want directly from someone they know and trust. Other markets include an individual building an oak log home and a business that makes "grave blankets" out of bigtooth aspen (used as a flat surface for placing flowers on graves). The highest quality red oak, white oak, hard maple, and cherry logs are sold for veneer to buyers that specialize in this market. Some lower-quality logs, mostly red maple, which is prone to defects and staining, is sold by the tractor trailer load to commercial mills that bid on and haul the timber from a landing on our property. I estimate that the premium prices garnered by Carl's marketing connections and expertise approximately compensate for the money that would have been paid for the additional trees that were marked and would have been sold had we accepted a bid from one of the mills. Those trees and many others that would likely have succumbed to the mechanical skidders are still growing in the forest.

4.1.7 Toward Grain and Oilseed Perennialism

A comprehensive understanding of all the ecological components of a place is not required in order to make good decisions about land management. The brief descriptions provided here of the diverse land that I care for are intended to provide concrete and imperfect examples for exploring the management challenges and objectives for a variety of places. Knowing enough about a place to be able to learn from and mimic its natural ecosystems is essential. When it is possible to allow natural ecological processes to restore degraded sites, the primary management task is to halt any ongoing damaging practices.† Halting overgrazing on moderately

---

* Burns and Honkala, *Silvics of North America*, vol. 2, *Hardwoods*.
† Dominick A. DellaSala et al., "A citizen's call for ecological forest restoration: forest restoration principles and criteria," *Ecological Restoration* 21, no. 1 (2003).

degraded grassland, for example, can be sufficient to reverse most of the damaging loss of soil organic carbon, nutrients, and productivity over time. Lengthening the time between timber harvests and selectively harvesting trees so that species and age diversity are maintained can result in improved ecological health in a forest ecosystem and preserve the timber resources over a longer time frame.[*] Causing as little harm as possible and preserving or enhancing diversity are universal principles of ecologically sound land management. The management challenge presented by the complexity of characteristics manifested by diverse tree species in an eastern deciduous forest, illustrated briefly in the Lower Michigan woodland example, illustrates that maintaining species richness sometimes requires varied practices from one acre or hectare to the next. Similarly, a pasture with an 8% slope will be prone to erosion from overgrazing before an adjacent field that is level, even if exposed to an identical stocking rate and duration. Simplicity and uniformity, central features of an industrial perspective, are not well suited to ecological restoration and maintenance.

In the current context of continued human population growth and scarce soil and water resources that profoundly threaten future food security for billions,[†] restoring healthy ecosystems is a necessary but not sufficient global land management strategy. The dilemma or central problem of agriculture has been succinctly framed by Wes Jackson as having two parts. The first is "how do we obtain an adequate, if not bountiful harvest?" The second is "how do we ensure that future adequate or bountiful harvests have not been compromised during production?"[‡] I believe this dilemma and the same questions apply equally to forest management. Indeed, the long-standing failure of cropping and grazing practices to produce an adequate harvest without compromising the capacity of the land to produce future crops has been and continues to be the primary cause of global deforestation as farmers and ranchers expand their operations to maintain or increase production as existing agricultural land is degraded.[§]

Jackson provides an excellent example of this process in Costa Rica that applies to tropical forests around the world. The traditional slash-and-burn farming system in Costa Rica has been changing to have a shortened rotation as food demand has increased. With less time between burning cycles, fewer nutrients accumulate in standing forests, so that when they are cleared and planted the crops have fewer nutrients available in the ash beds. A process that worked for thousands of years as "a cycle of restoration" has become "a cycle of degradation" and the result is declining yields with increased erosion and nutrient leaching.[¶] The cycle of degradation of land and water

---

[*] Maser, *Sustainable Forestry: Philosophy, Science, and Economics.*
[†] Food and Agriculture Organisation, *The State of the World's Land and Water Resources.*
[‡] Jackson, *Consulting the Genius of the Place,* 146.
[§] Helmut J. Geist and Eric F. Lambin, "Proximate causes and underlying driving forces of tropical deforestation," *BioScience* 52, no. 2 (2002).
[¶] Jackson, *Consulting the Genius of the Place,* 148.

resources is as old as agriculture and civilization. That cycle has accelerated exponentially during the age of fossil fuels and has recently been accelerated further by globalized industrialism. Exceeding the carbon sink capacity of the Earth's atmosphere and oceans is necessarily temporary. For civilization to meet the nutritional and other material needs of a growing population and preserve the finite functioning ecosystems upon which we all depend for survival, the long-standing cycle of degradation and pollution must be slowed and reversed. To do so will require creative and ecologically informed approaches to old and very large problems.

Wes Jackson and the Land Institute that he founded in the 1970s near Salina, Kansas, along with many cooperating scientists and farmers, have been developing an evolving vision and set of proposals for a radically (as in the Latin, *radic*, "at root") different approach to agriculture globally. The essence of the Jackson–Land Institute approach involves working with nature to replace annual grains grown in monocultures with perennial grains grown in polycultures. I urge the interested reader to consult Jackson's work directly to gain some sense of the philosophical, geological, and ecological contexts from which the pursuit of perennial grain crops emerges.[*] Jackson notes that "essentially all of nature's land-based ecosystems feature perennials," and this is true in "alpine meadows, tropical rain forests, desert scrub, deciduous forests and native prairies."[†] By developing an agriculture based in annual monocultures rather than perennial polycultures, humans simplified and disrupted ecological processes that had a long history of building and conserving soil, retaining and filtering water, and creating species richness and ecological diversity. Instead, we inadvertently created an agriculture that undermines all of these processes. Jackson's proposed solution will, like nature, take time.

Reestablishing diverse perennial vegetation in the food-producing regions of the globe would go a long way toward solving many of the ecological problems that we have been considering throughout this book. Soil degradation, water scarcity and pollution, chemical contamination of soil, and the nutrient runoff that is creating hypoxic dead zones where rivers meet the oceans could all be largely halted by returning cultivated land to perennial vegetation. But what would we eat? This is the fundamental problem taken up by Jackson, the Land Institute, and a growing consortium of collaborating scientists: "we humans obtain at least two-thirds of our total calories from grains and oilseed crops, none of them perennial. Existing perennial species can produce only a small fraction of the total calories required for direct consumption by a growing human population."[‡]

---

[*] Wes Jackson, *New Roots for Agriculture* (Lincoln: University of Nebraska Press, 1985); Wes Jackson, *Becoming Native to This Place: The Blazer Lectures for 1991* (Lexington: University Press of Kentucky, 1994); Jackson, *Consulting the Genius of the Place*; Wes Jackson, *Nature as Measure: The Selected Essays of Wes Jackson* (Berkeley, CA: Counterpoint Press, 2011).

[†] Jackson, *Consulting the Genius of the Place*, 151–152.

[‡] Jackson, *Consulting the Genius of the Place*, 154.

The Land Institute and researchers around the world are working to perennialize most of the major grain and oilseed crops, including wheat, grain sorghum, sunflower, rice, corn, and soybeans. Research is also proceeding to develop seed and grain crops from domesticated, more productive varieties of native perennial species such as Illinois bundleflower, intermediate wheatgrass, and eastern gamagrass. Chickpea, millets, flax, and a number of native plants also hold potential as perennial grain species. The Land Institute has developed a trademarked variety of intermediate wheatgrass that was first harvested in 2009. Kernza™ is expected to be ready for commercial grain production in about ten years. Jackson expects that it will take at least twenty-five years for researchers to develop a range of profitable, productive perennial grain crops. He and his colleagues have proposed a fifty year farm bill that has the goal of replacing 80% of U.S. annual grain acreage with perennials, with 20% remaining in annuals.[*]

Successful development and widespread adoption of perennial grain crops over the next fifty years will require developing new perennial plants through selective breeding, the creation of new agroecosystems utilizing multiple species in complementary polycultures, and relatively minor modifications of harvesting equipment. The anticipated benefits in comparison to annual cropping systems include greatly reduced runoff and soil erosion, resulting in more effective use of limited water resources and more nutrients available to crops with fewer nutrients in ocean ecosystems; perennial root systems grow deeper into the soil over multiple years and gain access to nutrients and moisture that is unavailable to annuals, resulting in greater drought resistance and fewer water and nutrient limitations on growth; perennial roots help to quickly build a granular soil structure that improves water absorption and nutrient availability; perennials develop a leaf canopy earlier in the spring than do annuals and continue photosynthesis and growth after harvest, extending the growing season and increasing the total solar energy converted to biomass; perennial mixes may also provide a superior habitat for nitrogen-fixing legumes, which will increase nitrogen availability and further reduce the need for nitrogen fertilizers.[†]

The coordinated effort by scientists to create productive perennial grain cropping systems as described by Wes Jackson is driven by a vision that has an ecological sensibility at its core:

> We can't go back to the crossroads where our ancestors took that wrong turn [in choosing annual monocultures], or to a golden age of folk agriculture that never existed. But we can now envision an agriculture in which we bring the ecological processes embodied within wild biodiversity to the farm, rather than forcing agriculture to relentlessly chip away at wild ecosystems.[‡]

---

[*] Jackson, *Consulting the Genius of the Place*, 156–187.
[†] Ibid, 161–163.
[‡] Jackson, *Consulting the Genius of the Place*, 152.

If perennial grains can be made progressively more available over the next ten–fifty years, much of the abandoned, degraded, and marginal cropland planted to high-diversity perennial grassland mixtures could be systematically brought out of grassland and into perennial crops. At the end of May 2012, 12 million hectares (29.60 million acres) were enrolled in the CRP in the United States.[*] This is primarily highly erodible former cropland planted to protective vegetation, most of it perennial native grass mixtures. Much of it would be suitable for perennial crops after at least ten years of building soil organic carbon.

The future acreage base for perennial grains is potentially much larger than the CRP acreage. Recent research identified 1.6 million hectares (4 million acres) of marginal cropland in Nebraska alone as appropriate for perennial biomass production. This study also tested a framework for identifying marginal croplands that includes agroeconomic profitability in combination with soil health indicators such as erosion, flooding, drainage, and slope.[†] This or a similar framework could be adapted and used globally to identify and prioritize land for LIHD grassland plantings and potential conversion to perennial crop production as perennial grains become viable. In the interim, these lands would be protected from further degradation by perennial grassland mixtures and also have the potential for biomass energy production while they are sequestering carbon and enhancing local ecological diversity and wildlife habitat.

4.1.8 Food Production with Existing Best Practices

Grain and oilseed crop production cannot be put on hold until perennial grain varieties are developed and made widely available. There has been a growing recognition in recent decades that producing grains in annual monocultures with intensive use of fossil fuel–based fertilizers and pesticides is causing profound soil degradation and a host of related ecological problems.[‡] Industrial "production" agriculture has been under attack from many quarters, mostly for environmental and health problems related to its products and production methods. A wide range of "alternative" agronomic approaches to growing food, from organic to permaculture to agroecology, are demonstrating that it is possible to produce good crop yields and reduce

---

[*] U.S. Department of Agriculture, "Conservation Reserve Program: Status-End of May 2012," ed. Farm Service Agency (2012) http://www.fsa.usda.gov/Internet/FSA_File/mayone pager2012.pdf.

[†] Gayathri Gopalakrishnan, M. Cristina Negri, and Seth W. Snyder, "A novel framework to classify marginal land for sustainable biomass feedstock production," *Journal of Environmental Quality* 40, no. 5 (2011).

[‡] Altieri, Agroecology; Food and Agriculture Organisation, *The State of the World's Land and Water Resources.*

the soil degradation and environmental damage produced by "conventional" industrial agriculture.*

The success of industrial food systems in producing cheap and plentiful animal protein and easily stored and shipped grain and oilseed commodities has led to a large shift in the human diet over much of the globe. This change has been, at its core, a shift away from eating the leaves of plants and away from eating animal products that come from animals fed mostly leaves of plants. Eating large quantities of animal products from animals fed large amounts of grain and eating increasing amounts of refined and processed grain products (along with sugar and sodium) is associated with affluence in Western societies. This diet has been exported to much of Asia, and the shift, wherever the "Western" diet has been adopted, has led to reduced intake of the omega 3 fatty acids plentiful in plant leaves and meat from animals that eat primarily plant leaves and toward an increased intake of omega 6 fatty acids, which are more plentiful in grain and animals fed grain. High rates of obesity, diabetes, heart disease, stroke, cancer, mental illness, and inflammatory disorders tend to accompany this dietary shift.[†]

Research designed to identify a healthy ratio of omega 3 and omega 6 fatty acids in the human diet and in the human body[‡] may provide an additional incentive to modify the nature of industrial agriculture as well as the diet associated with it. Eating less meat and feeding less grain to animals and at the same time eating more leaves has multiple potential benefits for the planet and human health.[§] The current practices of growing grain and oilseeds in annual monocultures on the Earth's most fertile land, using increasing portions of the crop produced to feed animals (many of which evolved to eat grass), and then diverting even more of the crop to produce fuel for automobiles is not a viable strategy for meeting increasing demand for food in an era of scarce and degraded soil and water resources.

Wes Jackson's vision and plan for an ecologically sound agriculture centered on perennial grains grown in polycultures may address many of the problems inherent to industrial agriculture in the future. That vision and plan, however, are not ready for immediate adoption. In the meantime, grains and oilseed crops will certainly continue to be grown and needed to meet the growing demand for food. There are proven methods of growing grain and oilseed crops that utilize both annual and perennial plants in mixtures and rotations. These methods are available now, are inexpensive, and avoid many of the ecological problems associated with industrial agriculture. One of the great advantages of production systems that utilize practices such as polycultures, cover cropping, mulching, crop rotation, and minimum tillage is their flexibility and adaptability to each unique place

---

* Altieri, *Agroecology*.
[†] Michael Pollan, *In Defense of Food: An Eater's Manifesto* (New York: Penguin Press, 2008).
[‡] Joseph R Hibbeln et al., "Healthy intakes of n–3 and n–6 fatty acids: estimations considering worldwide diversity," *American Journal of Clinical Nutrition* 83, no. 6 (2006).
[§] Pollan, *In Defense of Food*.

and circumstance.* They can be implemented incrementally by farmers and land owners interested in making moderate changes or who want to test the results of a changed practice before fully adopting it. A concrete example of one farmer's approach is provided here as an illustration of some "best practices" that can help to buy time and minimize damage as even better practices are developed.

Ed Reznicek raises beef cattle and grows corn, soybeans, cereal grains, and alfalfa on 162 hectares (400 acres) in northeastern Kansas. He has farmed the same land since 1978 using no pesticides or synthetic fertilizers. Ed is also active in the Kansas Organic Producers, a producers' marketing cooperative, and works part-time for the Kansas Rural Center, a nonprofit organization that promotes ecological agriculture and healthy rural communities. Most of Ed's family farm was owned by his wife Mary's family prior to Ed and Mary taking over the operation in the 1970s. Ed has a degree in philosophy and German, is a member of the local school board, and is active in his community. Ed grew up during the 1950s and 1960s on his parents' diversified family farm in east central Kansas. His family produced wheat, grain sorghum, corn, poultry, eggs, hogs, and beef cattle. He also did a short stint as an auto mechanic before going to college, where he developed skills that have proved useful on the farm.

Ed's work has been a mix of activism and farming, and the activism has been and remains closely tied to farming in ways that treat people, nature, and the land with respect. In my view, his farming practices approach the best that can be done, ecologically, until perennial grains are available. Ed is producing the most conventional and mainstream agricultural commodities (except for the legumes) using methods that are anything but conventional. His cropping system is centered on a complex and flexible crop rotation that features the extensive use of legumes and minimal tillage. He grows grain and oilseed crops on about 50–53 hectares (125–130 acres) of mostly terraced uplands. The remainder of the farm is pasture, hay ground, land used to produce grass seed, timber, grass waterways, and fence rows.

Ed initially became interested in organic farming in part as a way of reducing farm input costs. While working as an advocate for farmers during the farm credit and financial crises of the mid-1980s, he noticed that a large percentage of farmers' expenditures were for fuel, fertilizers, and pesticides. Replacing nitrogen fertilizers with legumes and eliminating pesticides while learning methods of nonchemical insect and weed control were ways to reduce costs and debt. Farming with used equipment and having the mechanical skills to keep it going also reduced expenses and avoided the energy costs embodied in new equipment.

At the center of Reznicek's farming operation is a "fairly disciplined crop rotation" over a seven-year period. He notes that the "key to success for organic crop rotation is to have enough legumes in the mix to produce the

---

* Altieri, *Agroecology*.

necessary nitrogen. The biannual and perennial legumes are also important for diversification of the cropping system and provide a foundation for effective nonchemical weed control." Alfalfa—"the queen of legumes"—is excellent forage for cattle, is a perennial, is a heavy nitrogen fixer, and is drought and disease resistant. The first year of Ed's rotation is a cereal grain, usually wheat or winter barley, planted in the fall. Alfalfa is drilled into the field of barley or wheat in the spring of the following year. After the wheat or barley are harvested in July for grain the remaining crop residue is cut and bailed for use as forage for cattle. Removing the straw and any weeds that have grown allows more sunlight to reach the alfalfa. The remainder of year 2 and year 3 of the rotation are devoted to alfalfa, which is cut and bailed for hay in three or four cuttings throughout the summer. In the spring of year 4, when the alfalfa is 30–38 centimeters (12–15 inches) tall, it is plowed into the ground as a green manure crop. A shallow plowing of 8–10 centimeters (3–4 inches) is used to cut the roots of the alfalfa. This allows some plants to survive and provides erosion control in the corn crop that is planted next. The remainder of the growing season the field is in corn. The fifth year the rotation goes to soybeans (also a legume), back to corn in year 6, and soybeans again in year 7. In five of the seven years each field is in nitrogen-fixing legumes and in three of the seven years there is no tillage. The rotation can be modified, for example, to include more cereal grains and less corn. The cereal grains have lower nutrient requirements than corn and can be used to control summer annual weeds.

The plantings of alfalfa also include red clover and burseem clover. The red clover is a biannual that overwinters well in northeast Kansas and the burseem clover is easily killed by freezing temperatures and functions as an annual in Ed's system. The burseem clover is planted for its early rapid growth and serves to suppress weeds in the cereal grains, in addition to its nitrogen contribution. The mix of legumes also helps to control the alfalfa weevil and provides diversity to ensure a stand of legumes even if there is an outbreak of alfalfa weevils. Significant damage from alfalfa weevils is rare due to a range of nonchemical controls. By not spraying pesticides, Ed believes that a population of insect predator species of the alfalfa weevil is maintained. The proximity of brushy fence rows and woody draws, a diversity of crops, and allowing cattle to graze the alfalfa residue during the fall and winter are other components of weevil control. The fencerows and woody draws provide habitat for insect predators. The cattle grazing and walking in the field reduces the number of weevil eggs that survive the winter in the stems of the alfalfa that remains.

Reznicek argues the Dust Bowl and environmental crises of the 1930s were partly caused by a widespread failure to use crop rotations. The value of crop rotations had been known and long advocated in USDA literature from that period, but they were never widely adopted in the United States. He notes that there is a popular myth that crop rotations were widely used prior to the availability of synthetic fertilizers and pesticides and that the need for rotations was diminished by the availability of these products. His system of crop rotation demonstrates that the opposite is true—synthetic fertilizers and chemical pesticides are unnecessary with a crop rotation that includes sufficient legumes and crop diversity. Ed's crop yields are comparable to those of his "conventional" neighbors who use fertilizers and pesticides to grow corn and soybeans and have not planted alfalfa for at least the past thirty–forty years.

About 12 hectares (30 acres) of the Reznicek farm are bottom ground and the remainder is in terraced upland fields. The areas between the terraces are managed as separate 1–2 hectare (3–5 acre) fields. The rotation of 25–30 separate fields on varying schedules adds considerably to the diversity of crops and legumes on the farm at any one time. It results in a production mix that includes cereal grains, corn, beans, and alfalfa harvested each year, adding economic diversity to the ecological mix. This also means that there is always alfalfa available for the cattle herd. A single seven-year rotation on the entire farm would increase economic risks if growing conditions were unfavorable for one of the crops in a year that it was the only one harvested.

The cattle distribute their manure on the fields when they graze alfalfa in the fall and winter. Additional manure is mechanically spread on the fields, usually between years 5 and 6 in the rotation before the second round of corn, which uses a lot of nutrients, and sometimes between years 7 and 1. Reznicek believes a widespread myth about organic farming is the belief that the majority of nitrogen is provided by manure. Manure is a good source of phosphorous, but only supplements the primary source of nitrogen, which is taken from the atmosphere and fixed in the soil by legumes. He notes that "we live in a pool of nitrogen, the atmosphere is 78% nitrogen and the only place that nitrogen is stable is in the atmosphere. When nitrogen fertilizer is put in the soil the nitrogen volatilizes into the atmosphere or leaches down through the soil." A variety of diverse legumes—perennials, biannuals, summer annuals, and winter annuals—are available for use in a variety of ways in different cropping systems. Legumes make buying nitrogen fertilizer unnecessary. "Nitrogen is all around us and we take it for granted."[*]

Reznicek's cropping system incorporates the central principles of ecologically sound land management and can be modified and adapted to a wide variety of conditions and circumstances. The system minimizes soil disturbance, enhances soil organic carbon and increases the species diversity of both plants and insects compared to the now very common industrial agriculture rotation between corn and soybeans using synthetic fertilizers and chemical pesticides. The use of legumes as green manure and as a cover crop and the integration of cattle and their manure into the cropping system eliminate the need to purchase fossil fuel–intensive

---

[*] Ed Reznicek, personal communication, January 10, 2012.

fertilizers. The legumes assist with weed control and the cattle reduce overwintering populations of alfalfa beetle, both of which help eliminate the need for chemical herbicides and insecticides. The grass waterways, brushy draws, and fencerows provide habitat for wildlife, including predatory insect species, which, in turn, also help to eliminate pest insects without chemicals. Each of these features also reduces wind and water erosion, and their proximity does not increase weed populations in crop fields.[*] The elimination of herbicides and insecticides helps to ensure a healthy mix of soil microorganisms and their complex interactions with fungi and organic matter. Water passing through the farm is not polluted with chemicals and nutrient runoff is minimal due to the terraces, grassy waterways, and the relative stability of nitrogen from legumes.[†] Each practice and component of a well-designed cropping system serves multiple interactive purposes. This multiplicity is shared by ecologically sound land management in forests and grasslands as well. The economic and social dimensions of ecological land management have similarly complex interactive features. Some of those features are introduced in the next section and are elaborated in Chapter 5.

4.1.9 Agroforestry and Reforestation in the Sahel

Deforestation has historically been severe in much of the sub-Saharan region of Africa known as the Sahel. Rapid deforestation accompanied by desertification continued in much of the region into the 1990s despite significant investments in reforestation projects by local, regional, and national governments and private development and relief organizations. Most of the reforestation projects attempted in the Sahel prior to the 1980s involved large-scale plantings of exotic species such as eucalyptus and neem. These projects required large central tree nurseries and were difficult to reproduce on a village level. As a result, the communities involved in the projects did not continue planting trees after the initial projects were completed. "Both government forestry services and non-government projects focused on the problem of deforestation and ignored the understanding of farmers."[‡]

A conservation and tree production system was started in 1983 in the Maradi region of Niger that has transformed Maradi, and much of Niger,

---

[*] H. Smith, L. G. Firbank, and D. W. Macdonald, "Uncropped edges of arable fields managed for biodiversity do not increase weed occurrence in adjacent crops," *Biological Conservation* 89, no. 1 (1999).

[†] Gary F. Zimmer, *The Biological Farmer : A Complete Guide to the Sustainable & Profitable Biological System of Farming*, 1st ed. (Austin, TX: Acres U.S.A., 2000).

[‡] P. J. Cunningham and T. Abasse, "Reforesting the Sahel: farmer managed natural regeneration," in *Domestication des espèces agroforestières au Sahel: situation actuelle et perspectives*, ed. A. Kalinganire, A. Niang, and A. Kone, Working Paper, World Agroforestry Centre, Nairobi, Kenya (2005).

and "has the potential to transform the entire Sahelian region."[*] The development of this system, known as Farmer Managed Natural Regeneration (FMNR), was begun with the recognition that Maradi farmers had been taught to consider trees as weeds that competed with crops. There had been laws in Niger, as in other African nations, that imposed fines for cutting trees, so farmers had learned to keep their land free of trees to avoid fines. A good farmer was one who had treeless land and kept his fields clear of any regrowth. After trees were removed the landscape was barren and infertile. Changing the situation required changing the way farmers viewed trees and the way they managed their land. The beginning of FMNR occurred when the local Maradi Integrated Development Project persuaded a few farmers to protect a certain percentage of the naturally regenerating native trees species on their farms for use as firewood. The trees most often used in FMNR are gao (*Faidherbia albida*) and baobab (*Adansanonia digitata*). The simple conservation technique of actively protecting and managing sprouts in order to re-create tree vegetation has evolved into a successful intensification of agricultural production at the same time that the environment has been improved.[†]

FMNR involves leaving one or more shoots on a portion of the long-lasting underground stumps of gao, baobab, and other trees that continue to produce shoots for years after being cut to the ground. Some of the shoots can be cut for firewood and some can be pruned and grown for timber products. The trees are grown in mixed agroforestry polycultures. Farmers cultivate grain and vegetable crops around the trees. The trees provide shade, protection from the wind, and help to retain moisture. Leaves, twigs, and small branches from the trees are placed on hardpan soil and are rapidly incorporated into the soil by termites. This improves soil structure, reduces soil erosion, and improves water infiltration. Wind deposits silt and fine organic matter as it passes over the woody debris and the fertility of previously unproductive soils is improved.[‡]

FMNR helps to improve crop yields and animal productivity. Grazing animals are drawn to edible tree pods, spend more time on the farms, and deposit manure, which improves fertility. The shade and mulch provided by the trees also help reduce high soil temperatures that can reduce productivity and volatilize nutrients in hot tropical and semiarid climates. The return of trees has been accompanied by the return of birds and predatory insects, which reduces the need for chemical pesticides. The income, quality of life, and environment have improved markedly in

---

[*] Cunningham and Abasse, "Reforesting the Sahel"; Ibid 2.

[†] M. Larwanou and C. Reij, "Farmer managed natural regeneration in Niger: a key to environmental stability, agricultural intensification, and diversification," in *Innovations as Key to the Green Revolution in Africa*, vol. 1, ed. Andre Bationo et al., (Dordrecht: Springer, 2011).

[‡] Cunningham and Abasse, "Reforesting the Sahel."

areas that have adopted FMNR.[*] Firewood, which is essential for cooking, is available for household use and as a cash crop. Timber is also used for both personal use and as a source of income. Grain and vegetable yields have increased. Reduced wind speeds and dust loads and increased shade have improved the human quality of life as the environment is more hospitable.[†]

FMNR is one important component of a broader set of institutional changes that have led to the reforestation and agricultural rejuvenation of more than 5 million hectares (12.4 million acres) in the Maradi and Zinder regions of Niger.[‡] Legal and governance changes were made to allow farmers to own and manage trees on the land they farm. The Maradi Integrated Development Program educated local farmers and initiated the project. Farmers shared their knowledge with kin and neighboring villages. In some areas the process had to be supplemented with planting seedlings because natural regeneration had stopped. The success of those practicing FMNR led to other methods of reforestation that were adapted to surrounding areas. The biophysical environment and economic viability of an entire region were enhanced by a diversity of actors, policies, and practices. This occurred in a region with one of the highest population growth rates on Earth and where long-term deforestation, ecological decline, famine, and poverty were endemic.[§]

FMNR requires no purchased inputs—no fertilizer, seed, or nursery stock—and no equipment except inexpensive manual pruning shears. The farmers that practice FMNR are not dependent on purchased inputs or outside experts once they have learned the basic system. FMNR meets the criteria noted earlier of a proven land management practice that is inexpensive, can be implemented immediately, and is reversible. Like Ed Reznicek's cropping rotation system and other ecological land management practices, each component and activity of FMNR results in multiple interrelated outcomes. Nurturing naturally regenerating native tree sprouts (perennials) leads to firewood, timber, livestock fodder, shade, windbreaks, improved soil organic matter and fertility, increased crop production, and improved income for the farmer. These changes together constitute a modest mitigation of global climate change in the form of increased carbon sequestration and a significant mitigation of long-term local and regional ecological degradation. The increased ecological resilience, drought tolerance, improved income, and more plentiful food supply also constitute a positive adaptation.

---

[*] Eric Haglund et al., "Dry land tree management for improved household livelihoods: farmer managed natural regeneration in Niger," *Journal of Environmental Management* 92, no. 7 (2011).
[†] Cunningham and Abasse, "Reforesting the Sahel."
[‡] Jan Sendzimir, Chris P. Reij, and Piotr Magnuszewski, "Rebuilding resilience in the Sahel: regreening in the Maradi and Zinder regions of Niger," *Ecology and Society* 16, no. 3 (2011), http://dx.doi.org/10.5751/ES-04198-160301.
[§] Sendzimir et al., "Rebuilding resilience in the Sahel."

4.2 Concrete Steps and a Vision

People in Niger, the Sahel, Africa, and all of us are faced with the necessity of attempting to mitigate and adapt to the most rapid warming of the planet since humans evolved. The increased heat in the Earth's atmosphere due to rapid increases in atmospheric concentrations of greenhouse gases (see Chapter 1) is accelerating the frequency of extreme weather events, droughts, and flooding. The people of the Sahel have been subjected to the persistent recurrence of drought and famine since the late 1960s.[*] The Sahel region may have experienced the direct effects of climate change as much or more than any heavily populated area on Earth. Only the sparsely populated poles and extreme high altitudes have seen greater changes.[†] The local adoption and sharing of a simple practice of nurturing the regeneration of native trees has made a key contribution to a highly successful reforestation and agricultural intensification in the Maradi and Zinder regions of Niger. This was done in the context of long-term deforestation, severe and recurrent drought, poverty, and rapid population growth.

As global climate and ecological conditions are deteriorating rapidly, examples of ecological resilience and restoration must be taken seriously and analyzed carefully for principles and practices that can be applied elsewhere. Each of the previous examples illustrates relatively inexpensive ways to enhance carbon biosequestration and increase ecological diversity. They all produce food, energy, or durable and valuable wood products in ways that are carbon neutral or carbon negative, or at the very least, with substantially lower carbon pollution than dominant industrial alternatives. The principles involved, although infinitely complex in their biological and chemical nature, can be adapted and applied in a range of different ecosystems. Restoring and protecting diverse mixtures of native perennial plants, or the closest approximation possible, in every biome and ecosystem will provide a foundation for the expansion of healthy ecosystems.

Land management and land use changes that reverse ecological degradation and transform the biophysical environment can produce outcomes that no other action, policy, or strategy can achieve. Reductions or even the elimination of all greenhouse gas emissions from fossil fuels, although desperately needed, cannot restore degraded grassland and cropland in time to restore and protect the soil needed to produce the food needed by 2050. Eliminating greenhouse gas emissions alone will not halt tropical deforestation or reverse the habitat destruction that is driving the dangerous loss of global ecological diversity. About 25% of current carbon emissions are

---

[*] Sendzimir et al., "Rebuilding resilience in the Sahel."
[†] Brian Dawson and Matt Spannagle, *The Complete Guide to Climate Change* (New York: Routledge, 2009).

due to tropical deforestation,[*] so even reducing carbon emissions requires land use and land management changes. Climate change is the aspect of global ecological decline that gets the headlines and emissions reductions are touted as the solution (or vilified as the end of capitalism and economic growth). Land use changes and land management practices are available now to mitigate and adapt to climate change and have a proven record of also directly addressing and improving a broad range of the ecological dimensions (including the economic and social) of the Earth's threatened habitability.

References

Allaby, M. 2005. *A Dictionary of Ecology*, 3rd ed. Oxford: Oxford University Press.
———. 2010. *A Dictionary of Ecology*, 4th ed. Oxford: Oxford University Press.
Altieri, M. A. 1995. *Agroecology: The Science of Sustainable Agriculture*, 2nd ed. Boulder, CO: Westview Press.
Bergquist, S. C. n.d. *The Glacial History and Development of Michigan*. East Lansing: State of Michigan.
Brown, L. R. 2011. *World on the Edge: How to Prevent Environmental and Economic Collapse*, 1st ed. New York: W. W. Norton.
Bruckerhoff, D. N. 2007. "Improving Your Woodland for Timber Production," L-725, Kansas State University Agricultural Experiment Station and Cooperative Extension Service, Manhattan, Kansas.
Burns, R. M., and B. H. Honkala. 1990. *Silvics of North America*, vol. 1, *Conifers*. Agriculture Handbook 654. Washington, DC: U.S. Forest Service.
———. 1990. *Silvics of North America*, vol. 2, *Hardwoods*. Agriculture Handbook 654. Washington, DC: U.S. Forest Service.
Cunningham, P. J., and T. Abasse. 2005. "Reforesting the Sahel: farmer managed natural regeneration." In *Domestication des espèces agroforestières au Sahel: situation actuelle et perspectives*, ed. A. Kalinganire, A. Niang, and A. Kone. Nairobi, Kenya: World Agroforestry Centre.
Dawson, B., and M. Spannagle. 2009. *The Complete Guide to Climate Change*. New York: Routledge.
DellaSala, D. A., A. Martin, R. Spivak, T. Schulke, B. Bird, M. Criley, C. van Daalen, J. Kreilick, R. Brown, and G. Aplet. 2003. "A citizen's call for ecological forest restoration: forest restoration principles and criteria." *Ecological Restoration* 21(1):14–23.
Doll, J. E., G. E. Brink, R. L. Cates, and R. D. Jackson. 2009. "Effects of native grass restoration management on above- and belowground pasture production and forage quality." *Journal of Sustainable Agriculture* 33(5):512–527.
Fargione, J., J. Hill, D. Tilman, S. Polasky, and P. Hawthorne. 2008. "Land clearing and the biofuel carbon debt." *Science* 319(5867):1235–1238.

---

[*] Pan et al., "A large and persistent carbon sink in the world's forests."

Field, C. B., J. E. Campbell, and D. B. Lobell. 2008. "Biomass energy: the scale of the potential resource." *Trends in Ecology and Evolution* 23(2):65–72.

Field, C. B., and M. R. Raupach. 2004. *The Global Carbon Cycle: Integrating Humans, Climate, and the Natural World*. Washington, DC: Island Press.

Food and Agriculture Organisation. 2011. *The State of the World's Land and Water Resources for Food and Agriculture: Summary Report*. Rome: United Nations.

———. "Carbon Content Estimation." http://www.fao.org/forestry/17111/en

Geist, H. J., and E. F. Lambin. 2002. "Proximate causes and underlying driving forces of tropical deforestation." *BioScience* 52(2):143–150.

Gillies, Carl. Personal communication. February 2, 2012.

Gopalakrishnan, G., M. C. Negri, and S. W. Snyder. 2011. "A novel framework to classify marginal land for sustainable biomass feedstock production." *Journal of Environmental Quality* 40(5):1593–1600.

Haglund, E., J. Ndjeunga, L. Snook, and D. Pasternak. 2011. "Dry land tree management for improved household livelihoods: farmer managed natural regeneration in Niger." *Journal of Environmental Management* 92(7):1696–1705.

Hibbeln, J. R., L. R. G. Nieminen, T. L Blasbalg, J. A. Riggs, and W. E. M. Lands. 2006. "Healthy intakes of n–3 and n–6 fatty acids: estimations considering worldwide diversity." *American Journal of Clinical Nutrition* 83(6):S1483–S1493.

Hong, B. D., and E. R. Slatick. 1994. "Carbon dioxide factors for coal." *Quarterly Coal Report* January–April, http://www.eia.gov/cneaf/coal/quarterly/co2_article/co2.html

Jackson, T. 2011. *Prosperity Without Growth: Economics for a Finite Planet*. London: Earthscan.

Jackson, W. 1985. *New Roots for Agriculture*. Lincoln: University of Nebraska Press.

———. 1994. *Becoming Native to This Place: The Blazer Lectures for 1991*. Lexington: University Press of Kentucky.

———. 2010. *Consulting the Genius of the Place: An Ecological Approach to a New Agriculture*. Berkeley, CA: Counterpoint Press.

———. 2011. *Nature as Measure: The Selected Essays of Wes Jackson*. Berkeley, CA: Counterpoint Press.

Kansas Forest Service. 2011. "Loss of forests: development or timber harvest." *Kansas Canopy*, Winter (41):10–11.

Lal, R., R. F. Follett, B. A. Stewart, and J. M. Kimble. 2007. "Soil carbon sequestration to mitigate climate change and advance food security." *Soil Science* 172(12):943–956.

Larwanou, M., and C. Reij. 2011. "Farmer managed natural regeneration in Niger: a key to environmental stability, agricultural intensification, and diversification." In *Innovations as Key to the Green Revolution in Africa*, vol. 1, ed. A. Bationo, B. Waswa, J. M. M. Okeyo, F. Maina, and J. M. Kihara, 1311–1319. Dordrecht, The Netherlands: Springer.

Little, E. L., ed. 1998. *National Audubon Society Field Guide to North American Trees: Eastern Region*. New York: Alfred A. Knopf.

Lyle, S. A. 2009. *Glaciers in Kansas*. Lawrence: Kansas Geological Survey.

Maestre, F. T., J. L. Quero, N. J. Gotelli, A. Escudero, V. Ochoa, M. Delgado-Baquerizo, M. García-Gómez, et al. 2012. "Plant species richness and ecosystem multifunctionality in global drylands." *Science* 335(6065):214–218.

Maser, C. 1994. *Sustainable Forestry: Philosophy, Science, and Economics*. Boca Raton, FL: St. Lucie Press.

McPherson, B. J., and E. T. Sundquist. 2009. *Carbon Sequestration and Its Role in the Global Carbon Cycle*. Washington, DC: American Geophysical Union.

Pan, Y., R. A. Birdsey, J. Fang, R. Houghton, P. E. Kauppi, W. A. Kurz, O. L. Phillips, et al. 2011. "A large and persistent carbon sink in the world's forests." *Science* 333(6045):988–993.

Pollan, M. 2008. *In Defense of Food: An Eater's Manifesto*. New York: Penguin Press.

Raloff, J. 2011. "Environment: chemicals linked to kids' lower IQs: studies identify effects from pesticides still used on farms." *Science News* 179(11):15.

Reznicek, Ed. Personal communication. January 10, 2012.

Russelle, M. P., R. V. Morey, J. M. Baker, P. M. Porter, and H.-J. G. Jung. 2007. "Comment on 'Carbon-negative biofuels from low-input high-diversity grassland biomass.'" *Science* 316(5831):1567.

Salatin, J. 1995. *Salad Bar Beef*, 1st ed. Swoope, VA: Polyface.

Sendzimir, J., C. P. Reij, and P. Magnuszewski. 2011. "Rebuilding resilience in the Sahel: regreening in the Maradi and Zinder regions of Niger." *Ecology and Society* 16(3):1, http://dx.doi.org/10.5751/ES-04198-160301.

Sims, R. A., and K. A. Baldwin. 1991. *Landform Features in Northwestern Ontario*. COFDRA report 3312. Ontario, Canada: Ontario Ministry of Natural Resources.

Smith, H., L. G. Firbank, and D. W. Macdonald. 1999. "Uncropped edges of arable fields managed for biodiversity do not increase weed occurrence in adjacent crops." *Biological Conservation* 89(1):107–111.

Socolow, R. H. 2005. "Can we bury global warming?" *Scientific American* 293(1):49–55.

Stokstad, E. 2011. "Can biotech and organic farmers get along?" *Science* 332(6026):166–169.

———. 2012. "Engineered superbugs boost hopes of turning seaweed into fuel." *Science* 335(6066):273.

Tilman, D., J. Hill, and C. Lehman. 2006. "Carbon-negative biofuels from low-input high-diversity grassland biomass." *Science* 314(5805):1598–1600.

U.S. Department of Agriculture. 2011. "Conservation Reserve Program: Status-End of December 2011." Farm Service Agency.

———. 2011. "Map Unit Description, Soils Inventory Report." Natural Resources Conservation Service. Jefferson County, Kansas.

Volland, C. 2011. "Flint Hills smoke management plan off to a shaky start." *Planet Kansas*.

Wargacki, A. J., E. Leonard, M. N. Win, D. D. Regitsky, C. N. S. Santos, P. B. Kim, S. R. Cooper, et al. 2012. "An engineered microbial platform for direct biofuel production from brown macroalgae." *Science* 335(6066):308–313.

Williams, J. H., A. DeBenedictis, R. Ghanadan, A. Mahone, J. Moore, W. R. Morrow, S. Price, and M. S. Torn. 2012. "The technology path to deep greenhouse gas emissions cuts by 2050: the pivotal role of electricity." *Science* 335(6064):53–59.

www.fueleconomy.gov. "How can a gallon of gasoline produce 20 pounds of carbon dioxide?" http://www.fueleconomy.gov/feg/co2.shtml

Zimmer, G. F. 2000. *The Biological Farmer: A Complete Guide to the Sustainable & Profitable Biological System of Farming*, 1st ed. Austin, TX: Acres U.S.A.

5

Conservation Policy and the Politics of Growth

5.1 Conservation Assistance Is Available

Each of the practices discussed in Chapter 4 and others that incorporate the principles of perennialism, minimal disturbance, and diversity can be adapted and implemented in a wide range of local ecosystems. In the United States, for example, technical assistance and financial resources for conservation and ecological restoration are available through several agencies of the U.S. Department of Agriculture (USDA) and numerous state agencies, local boards, and conservation districts. The USDA's Natural Resources Conservation Service (NRCS) and Farm Service Agency (FSA) serve as accessible clearinghouses for information about conservation practices and programs. Information is available to anyone interested and can be obtained at local FSA and NRCS offices in every state and online.[*] In Canada, the Ministry of Natural Resources in each province provides access to information about conservation programs, technical assistance, and resources.[†] Information about and access to public conservation resources can be obtained from comparable departments and ministries in each nation. Protecting and enhancing public resources for natural resource conservation and improvement at local, regional, state, and international levels is the policy equivalent of improved land management practices; it is relatively inexpensive and can be utilized immediately because proven programs are in place.

I have found NRCS and FSA personnel to be helpful and generally knowledgeable about the programs under their jurisdiction as well as state and local resources and federal programs that they do not administer. They often work closely with local county extension agents, for example, as well as state forest service and agriculture departments. In Kansas and most other states,

---

[*] U.S. Department of Agriculture, "NRCS Conservation Programs," http://www.nrcs.usda.gov/programs/; U.S. Department of Agriculture, "Conservation Programs," http://www.fsa.usda.gov/conservation

[†] Ontario Ministry of Natural Resources, " Ontario Ministry of Natural Resources," http://www.mnr.gov.on.ca/en/

nonprofit organizations, like the Kansas Rural Center, work directly with NRCS to train personnel on issues ranging from organic farming practices to the use of seasonal high tunnels (greenhouses) and to communicate the resource needs of farmers, ranchers, and forest managers. The financial resources available for conservation practices are limited, but can defray some or all of the costs of implementing ecologically restorative practices. I do not always agree with or follow the recommendations of USDA staff, but I have found a remarkable degree of flexibility in program administration. My experience with the programs that I participate in will illustrate this point and provide concrete examples of the kinds of practices for which technical and financial assistance are available.

I currently participate in three USDA conservation programs: the Conservation Reserve Program (CRP), the Biomass Crop Assistance Program (BCAP), and the Environmental Quality Incentives Program (EQIP). Our annual payments for all three programs total $6,772. Over the five-year term of the BCAP contract there are additional payments of up to about $2,000 for one-time reimbursement of expenses related to establishment of the biomass grass stand and up to about $2,400 per year for two years as a matching payment for the first two biomass crops sold under BCAP.

The CRP contract pays "rent" on 20.6 hectares (51 acres) of highly erodible cropland that has been planted to perennial grasses on our Michigan farm. This cropland was previously used for growing corn and beans by a neighbor who used synthetic fertilizers, chemical herbicides, and pesticides. This land drains partially into wetlands on the property that are no longer being polluted by chemicals, nutrients, and soil sediments since the conversion to perennial grasses under the CRP. The mixed perennial grasses sequester more carbon in the soil than the crops they replaced, reduce erosion, increase plant species diversity, and provide high-quality habitat for a wide variety of wildlife. The fossil fuels used to control weeds and brush by spot mowing each year is a fraction of the fuel used for plowing, disking, planting, spraying, and harvesting corn and beans. The soil has not been disturbed for ten years, which has reduced nitrogen oxide and other soil emissions. This CRP contract runs for a period of fifteen years, which ends in 2017. The contract is transferable if the land changes ownership. It likely can be renewed when the contract expires (if congressional funding for the program continues) or this land could be returned to crop production. If the land is returned to grain and oilseed crop production it will have a higher organic carbon content and water retention capacity than when it was first planted to mixed grasses and will be surrounded by an established vegetative buffer to minimize future runoff.

The planting of perennial vegetation on CRP land is inexpensive, available now, and reversible. The program can provide rent and cost-share resources for the establishment and maintenance of grass or trees and shrubs on erodible cropland. There were about 12 million hectares (29.60 million acres) enrolled nationally in the CRP program in May 2012, down from about

14.16 million hectares (35 million acres) a few years earlier.[*] This program, in place since 1985, has likely done more to conserve soil, improve water quality, sequester carbon, and increase ecological diversity on U.S. agricultural land than any program in history. It serves as a model public policy for enhancing biosequestration of carbon, ecological restoration, and future food security. High commodity prices, especially for subsidized corn, can be a strong incentive for landowners to put land back into crop production as CRP contracts expire. The acreage in CRP may continue its recent decline in 2012 and coming years. Policy incentives to retain CRP land in perennial vegetation and adequate funding to offer future contracts make economic and ecological sense.

Our BCAP contract, as described in Chapter 4, pays annual rent on 10.5 hectares (26 acres) of grassland (most of which was former cropland) on our Creekridge Farm in Kansas. BCAP is intended to provide incentives to landowners for the establishment and production of (nongrain) biomass crops for conversion to energy and advanced biofuels. BCAP can be used on forestland as well as grassland. The program pays, in our case, to establish a mixed stand of warm season native grasses and legumes to be harvested for biomass energy. BCAP pays an annual per-acre rent for a five-year period, 75% of the one-time costs for ground preparation, seed stock, and planting expenses, and a matching payment of up to $45 per ton for the first two biomass crops harvested on enrolled acres. In our case, 5.25 hectares (12.5 acres) of the 10.5 hectares (26 acres) were already in established native grasses, which reduced overall establishment costs. A district conservationist at NRCS in our area agreed to allow a more diverse planting than the five varieties of grass and legumes recommended in NRCS guidelines. This adjustment was made after I provided information that higher-diversity plantings will produce more biomass and have several additional soil and ecological benefits.[†] I planted four additional grass species and a mix of twenty native flowers and forbs (broadleaf herbs), at my own expense, to add species diversity.

BCAP Project Area 1, which includes thirty-nine counties in western Missouri and eastern Kansas, enrolled 8,094 hectares (20,000 acres) in 2011, with an eventual goal for the area of 20,235 hectares (50,000 acres). The program was created in the 2008 Farm Bill and 2011 was the first year of enrollment, so its eventual national extent is not yet known. New local BCAP areas are being established (in 2012 and beyond) and administered by the FSA. The future extent of expansion will likely be largely constrained by available funding. The budget for the program in 2012 is $17 million and a total of 20,197 hectares (49,908 acres) had been enrolled as of March 2012.[‡]

---

[*] U.S. Department of Agriculture, "Conservation Reserve Program: Status-End of May 2012," ed. Farm Service Agency (2011).
[†] David Tilman, Jason Hill, and Clarence Lehman, "Carbon-negative biofuels from low-input high-diversity grassland biomass," *Science* 314, no. 5805 (2006).
[‡] Kelly Novak, personal communication, Feb. 10, 2012.

The BCAP is an important step toward establishing reliable local sources of biomass, an essential requirement for the future development of economically viable and ecologically sound liquid or gaseous cellulosic fuels. As the processes and facilities for advanced biofuels are in development, current biomass production is used to create compressed biomass pellets that can be burned for heat or electricity generation to displace fossil fuels. The carbon biosequestration and other ecological benefits, low cost, and potential reversibility back to cropland are similar to the CRP. The five-year contract period for BCAP is shorter than the ten-year and fifteen-year CRP contracts. This provides more flexibility for the landowner and should allow enough time to assess the economic viability of cellulosic biomass as a source of heat, electricity generation, and advanced biofuels. The BCAP does not have the proven history of the CRP but may prove to provide sufficient incentive for an acceleration of the development of cellulosic biofuels. Such development is a small but essential component of meeting the increasingly pervasive public policy goal of reducing greenhouse gas emissions by 80% below 1990 levels by 2050.[*]

The EQIP is a broad USDA program that provides technical and financial assistance for conservation practices on agricultural and forestland. In 2011 more than 5.3 million hectares (13 million acres) were enrolled on more than 38,000 tracts of land. EQIP provides assistance for projects ranging from air quality and on-farm energy to water quality and forestland health. EQIP also provides assistance to farmers making the transition to organic crop production and to those wanting to extend the growing season for fruits and vegetables with the use of seasonal high tunnels (a form of greenhouse). Financial assistance commitments to landowners enrolled in EQIP totaled about $865 million in 2011.[†] Total budget authority for the USDA in 2011 was about $149 billion, with 7%, or about $10 billion, budgeted for all conservation and forestry programs. Nutrition programs accounted for 70% of the budget, and farm and commodity programs accounted for 17%.[‡]

Our Creekridge EQIP contract is to implement timber stand improvement practices to improve forest health on 3.5 hectares (8.6 acres) of woodland over a three-year period. The NRCS administers the program with participation of the Kansas Forest Service. We have had a long-term relationship with the Forest Service to develop and implement a Forest Stewardship Plan with technical assistance from a district forester. The EQIP Forest Health contract provides resources to implement that plan with ongoing technical assistance and oversight from a professional forester. The financial assistance totals

---

[*] James H. Williams et al., "The technology path to deep greenhouse gas emissions cuts by 2050: the pivotal role of electricity," *Science* 335, no. 6064 (2012).

[†] U.S. Department of Agriculture, "Environmental Quality Incentives Program," http://www.nrcs.usda.gov/wps/portal/nrcs/main/national/programs/financial/eqip

[‡] U.S. Department of Agriculture, "FY 2011 Budget Summary and Annual Performance Plan," http://www.obpa.usda.gov/budsum/FY11budsum.pdf

$2,519 over three-years in our case and is intended to improve the health and future timber productivity of the forestland. We enrolled the portion of our 9.3 hectares (23 acres) of eastern deciduous forest that has the highest concentration of commercially valuable tree species, mostly black walnut and bur oak. Activities include removing invasive species, pruning and opening the canopy for future crop trees, and planting young trees in areas where natural regeneration is weak. This work takes many hours of labor per acre. The payment of about $729 per hectare ($295 per acre) is made after the work is complete and reviewed by the participating forester. This amount will pay for any needed tools and seedlings and possibly a few dollars per hour for labor. The collective public benefit is a healthier stand of forest and the air quality, water retention, and carbon biosequestration improvements associated with healthy forest ecosystems.

The USDA funds and administers several other conservation programs. The Conservation Technical Assistance program provides a wide range of technical assistance to help landowners and managers conserve and improve natural resources. There is a Wildlife Habitat Incentives Program, Wetland Reserve Program, and incentives under CRP to establish riparian buffers that protect streams and lakes from chemical and sediment pollution flowing from cropland. The Conservation Stewardship Program, available in all fifty states and the Pacific and Caribbean areas, is focused broadly on improving soil and water resources and conservation practices that address the effects of climate change.[*] The publicly funded conservation programs and resources briefly summarized here, and many more created and administered by states and local governments, are available to assist with ecological restoration in cropland, grassland, forests, wetlands, and urban environments—the bulk of the terrestrial ecosphere. As long as grain and oilseed crops are annual plants, agriculture will continue—more or less depending on methods and topography—to be a dangerous source of soil degradation and erosion.[†] Publicly funded conservation practices are an essential ecological defense and, in part, a bridge to a time when perennialism can expand from forests and grasslands to cropland. There is also evidence that broad conservation policies like those briefly described above are at least as effective at maximizing carbon sequestration and species diversity as policies with highly targeted incentives for those objectives.[‡]

---

[*] Kansas Association of Conservation Districts, *Conservation in Kansas: An Overview* (Lawrence: Kansas Association of Conservation Districts, 2012).

[†] Wes Jackson, *Consulting the Genius of the Place: An Ecological Approach to a New Agriculture* (Berkeley, CA: Counterpoint Press, 2010).

[‡] Erik Nelson et al., "Efficiency of incentives to jointly increase carbon sequestration and species conservation on a landscape," *Proceedings of the National Academy of Sciences of the United States of America* 105, no. 28 (2008).

5.2 Societal Transformation and the Politics of Growth

5.2.1 From Degradation to Restoration

Changing the ways we grow food, manage forests, and nurture and protect grasslands and other terrestrial ecosystems are clearly viable ways to enhance the biosequestration of carbon, reduce atmospheric concentrations of carbon dioxide (CO_2), and slow ecological degradation. They are also components of a structural transformation of industrial capitalist economies. Protecting and enhancing terrestrial and marine ecosystems, coupled with the drastic reduction in fossil fuel emissions that is required to salvage a habitable planet, will require a set of changes so profound that what we know as industrialism and capitalism must be largely transformed. Industrial economies have become so dependent on fossil fuels, unending growth of material throughput, and degradation of the environment that these features define their essence.[*]

Creating economies and societies that use far less energy, utilize primarily diffuse forms of renewable energy, reduce material throughput, and restore ecological health will require changes so fundamental that the resulting economic and social structures and processes will bear little resemblance to those of current industrial capitalist societies.[†] This observation should not be mistaken as a call for the end of representative democracy or the replacement of a mixed economy that has both private and public ownership of land and productive assets. On the contrary, both of those systems can and generally have worked better than any known alternatives.[‡] However, a transition to ecologically sound economies and societies will require a structural transformation that goes to the heart of contemporary industrialism. Globalized industrial capitalism is driven by the imperative for economic growth.[§] Finite ecosystems are incompatible with unending growth of natural resource consumption. This is a matter of basic physics and logic.[¶] The transition to social and economic forms that meet the needs of a growing population while reducing material throughput and waste, however, presents a complex and perplexing dilemma.[**]

A transition away from fossil fuels toward carbon-neutral and carbon-negative energy sources and infrastructure will require decades of research,

---

[*] William A. Faunce, *Problems of an Industrial Society*, 2nd ed. (New York: McGraw-Hill, 1981).

[†] Larry Lohmann, "Capital and climate change," *Development and Change* 42, no. 2 (2011).

[‡] John Michael Greer, *The Wealth of Nature: Economics as if Survival Mattered* (Gabriola, BC, Canada: New Society Publishers, 2011).

[§] Robert J. Antonio, "Climate change, the resource crunch, and the global growth imperative," in *Current Perspectives in Social Theory*, vol. 26, ed. Harry F. Dahms (Bingley, UK: Emerald Group Publishing, 2009).

[¶] Herman E. Daly, *Beyond Growth: The Economics of Sustainable Development* (Boston: Beacon Press, 1996).

[**] Tim Jackson, *Prosperity Without Growth: Economics for a Finite Planet* (London: Earthscan, 2011).

technology development, and private and public investment. This process has begun and must continue to accelerate if the policy goals for emissions reductions that are emerging at all levels of government are to be met.* At the same time, a transition has also begun away from land use and land management practices that are now known to produce near universal ecosystem degradation and contribute to carbon emissions. This second transition has an advantage over the first in that it can be expanded quickly and at much less expense.†

The societal and economic transformation embodied in improved land use and land management has been initiated and its methods and practices are being implemented and honed in forests and fields on every continent and in every nation. Tropical deforestation, although still a profound threat to livelihoods and the global carbon cycle, has finally slowed in the last decade.‡ Farmers who forgo synthetic nitrogen fertilizer to use legumes and manure take a small step toward the creation of a revolutionary new food system that will no longer be dependent on fossil fuels. Locally produced liquid fuels from terrestrial or algal biomass will assist with another important step in that process. The development of perennial grain and oilseed crops has been initiated and may end the annual soil disturbance that has degraded soils since the emergence of agriculture 10,000 years ago.§

Small landholders in Niger who nurture the regeneration of native trees to create an agroforestry system that produces more food, conserves scarce water, and restores carbon to the soil are successfully adapting to climate change and enhancing biological diversity. At the same time they are demonstrating the viability of agricultural intensification on existing cropland that the United Nations (UN) Food and Agriculture Organisation (FAO) has identified as essential to future food security.¶ Landowners and farmers who restore diverse perennial native grassland plants, plant perennial food-producing shrubs and trees in their yards and fields, reduce or eliminate mowing with fossil-fueled machines, stop watering their lawns and gardens with chlorinated water from municipal water plants, compost yard and kitchen wastes, and grow food in soil improved by their actions are not simply making "lifestyle" choices. They are engaging in land management practices that have structural implications, changing the nature of

---

* Williams et al., "The technology path to deep greenhouse gas emissions cuts by 2050."
† Wilfred M. Post et al., "Terrestrial biological carbon sequestration: science for enhancement and implementation," in *Carbon Sequestration and Its Role in the Global Carbon Cycle*, ed. Brian J. McPherson and Eric T. Sundquist (Washington, DC: American Geophysical Union, 2009).
‡ Yude Pan et al., "A large and persistent carbon sink in the world's forests," *Science* 333, no. 6045 (2011).
§ Jackson, *Consulting the Genius of the Place.*
¶ Food and Agriculture Organisation, *The State of the World's Land and Water Resources for Food and Agriculture: Summary Report* (Rome: United Nations, 2011).

their relationship to land and the atmosphere from one of degradation to one of restoration.

The extent to which these and similar actions transform societies away from ecological degradation and intensified climate change depends more on the extent of their adoption (social action) than on the creation of new practices or complex technologies. Although new technologies and their adoption will be essential, the required transformation is social and political, and involves a fundamentally changed relationship between humans and nature. Changing the way governments and corporations use energy and land will be required, but those who can take concrete actions need not wait and are not waiting for this to occur.[*]

It is also not necessary to depend on the next supposed great technological savior—be it genetically modified foods or carbon capture and storage (CSS)—to begin implementing well-established principles and practices of restorative land use and management, energy conservation, or adoption of low- or no-carbon energy sources. What is required first is recognition of the nature and urgency of the problem and a commitment to act personally and politically. This does not mean the problem of climate change and associated global ecological degradation can be solved. The damage to the carbon cycle, anthropogenic climate change, and ecological degradation are long-lasting realities to which we must adapt and begin to reverse by treating land and using energy differently. Doing so will require collective actions halting the "greatly accelerated…speed and volume of the throughput of natural resources and creation of wastes" created by the recent globalization of industrial capitalism.[†]

5.2.2 A Sense of Urgency

Many of the scientists with the most intimate scientific and personal understanding of the emerging climate and ecological catastrophe on Earth have been making increasingly urgent pleas for immediate action. Geoscientists with the American Geophysical Union have embarked on a comprehensive interdisciplinary effort to identify ways to stabilize the global carbon cycle in order to preserve the "habitability" of the planet.[‡] Tim Flannery, a renowned Australian scientist and author, provided an extensive and detailed explanation of climate science and warned of a precarious future when he wrote *The Weather Makers: How Man is Changing the Climate and What it Means for Life on*

---

[*] Kolya Abramsky, *Sparking a Worldwide Energy Revolution: Social Struggles in the Transition to a Post-Petrol World* (Oakland, CA: AK Press, 2009).

[†] Antonio, "Climate change, the resource crunch, and the global growth imperative."

[‡] Steven W. Running et al., "Next-generation terrestrial carbon monitoring," in *Carbon Sequestration and Its Role in the Global Carbon Cycle*, ed. Brian J. McPherson and Eric T. Sundquist (Washington, DC: American Geophysical Union, 2009).

Earth, published in 2005.[*] Flannery then reassessed the newly emerging scientific evidence that was published subsequent to the 2007 Intergovernmental Panel on Climate Change (IPCC) report (AR4), publishing *Now or Never: Why We Must Act Now to End Climate Change and Create a Sustainable Future* in 2009.[†] In this book Flannery shortened his review of the science and urged immediate action on multiple fronts, even to the point of urging that we be prepared to resort to dangerous geoengineering schemes like injecting sulfur dioxide into the upper atmosphere to reflect solar energy away from the Earth; certainly not the preferred choice of an ecologist. But by offering drastic measures as an option Flannery drove home his point—we are facing a life-threatening emergency. Flannery also identified land use and land management practices as immediately available means to begin to reverse dangerous trends. The more extensive our adoption of such practices, the less likely we will be forced to resort to dangerous, expensive, unproven, and possibly irreversible emergency geoengineering.

The urgency of National Aeronautics and Space Administration (NASA) scientist James Hanson's 2009 book *Storms of My Grandchildren* is clear in the subtitle, *The Truth About the Coming Climate Catastrophe and Our Last Chance to Save Humanity*. Hanson provides a succinct summary of our collective dilemma and the actions he views as needed to change course:

> It's crucial that we immediately recognize the need to reduce atmospheric carbon dioxide to at most 350 ppm in order to avoid disasters for coming generations. Such a reduction is still practical, but just barely. It requires a prompt phaseout of coal emissions, plus improved forestry and agricultural practices....we need to acknowledge now that a change of direction is urgent. This is our last chance.[‡]

Hanson's emphasis on phasing out coal and utilizing improved forestry and agricultural practices reflects his recognition that these are the actions most likely to have rapid positive effects on atmospheric concentrations of CO_2. Coal is technically easier to replace than liquid fuels from petroleum and is the dirtiest of the fossil fuels in terms of aerosols, toxic pollutants, and human health impacts. Improved forestry and agricultural practices are available immediately, are inexpensive, and have numerous ecological benefits.[§]

The eighteen past winners of the unofficial Nobel for the environment—the Blue Planet prize—were brought together in February 2012 to commemorate

---

[*] Tim F. Flannery, *The Weather Makers: How Man Is Changing the Climate and What It Means for Life on Earth*, 1st U.S. ed. (New York: Atlantic Monthly Press, 2005).

[†] Tim F. Flannery, *Now or Never: Why We Must Act Now to End Climate Change and Create a Sustainable Future*, 1st ed. (New York: Atlantic Monthly Press, 2009).

[‡] James E. Hansen, *Storms of My Grandchildren: The Truth About the Coming Climate Catastrophe and Our Last Chance to Save Humanity*, 1st U.S. ed. (New York: Bloomsbury USA, 2009), ix–x.

[§] Hansen, *Storms of My Grandchildren*.

the fortieth anniversary of the founding of the UN Environment Program. These leading scientists were asked to collectively author a statement to be presented to the Rio +20 Earth Summit Conference in June 2012. They characterized our current situation as an "absolutely unprecedented emergency" marked by a "rapidly deteriorating biophysical situation" that is "barely recognized by a global society infected by the irrational belief that physical economies can grow forever." The belief that economies can grow perpetually and that economic growth will cure all the world's problems was identified by this group of scientists as "the root cause of our unsustainable global practices."[*]

The scientists cited above and those referenced in earlier chapters are in agreement that immediate, urgent, and concerted action is required to avoid increasingly catastrophic effects from anthropogenic climate change and related global ecological degradation. The emphasis on rapidly implementing improved forestry and agricultural practices, as well as phasing out coal, the dirtiest and most easily replaced of the fossil fuels, reflects, in part, an understanding that the near total elimination (or decarbonization) of fossil fuels is a difficult and time-consuming process. A short summary of the complex changes and developments that must occur for one version of that energy transformation to be accomplished is provided below. The version summarized is the dominant one in industrial societies—it assumes continued growth of material consumption and no substantial lifestyle changes. Although these assumptions ignore the reality of finite ecosystems,[†] they continue to permeate most economic and policy proposals regarding climate change.[‡]

5.2.3 Changing the Energy System and Infrastructure

Transforming industrial capitalism to social forms that slow and reverse anthropogenic climate change and global ecological degradation will require profound changes in energy use, sources, and systems and cannot be ignored as an essential component of ecological restoration. Both energy use and land management must change to maintain viable, healthy ecosystems. The reader interested in learning more about the many ongoing grassroots democratic efforts to transform local and global energy systems and create renewable and more democratic alternatives is encouraged to read the works cited in this chapter, especially the work of Kolya Abramsky.[§]

In a 2010 article published in *Energy Policy*, Johannes Bollen and his colleagues argued that policies directed toward energy security, climate change, and pollution control will have the best outcomes if there is close

---

[*] John Vidal, "Civilisation faces 'perfect storm of ecological and social problems.'" *The Guardian*, February 20, 2012, 1–2, http://www.guardian.co.uk/environment/2012/feb/20/climate-change-overconsumption
[†] Daly, *Beyond Growth*.
[‡] Jackson, *Prosperity Without Growth*.
[§] Abramsky, *Sparking a Worldwide Energy Revolution*.

coordination and integration of all three policy areas. Extending scarce petroleum supplies for essential uses, reducing premature deaths from air pollution by up to 3 million people per year globally, and constraining the global average atmospheric temperature increase to 3°C above preindustrial levels were identified as realistic goals with well-integrated energy policies.[*] Note, once again, that yet another recent scientific assessment has concluded that dangerous climate change is very likely in our future despite our best efforts. Mitigating the rise in temperature and related climate and ecological changes as much as possible and adapting to the current and coming changes with new energy systems, lower consumption, and available land management practices are our central remaining options.

A look at what is thought to be required by scientists and engineers specializing in energy systems to accomplish the near total elimination of fossil fuels is sobering and helpful in developing concrete plans for the immediate future. Many states and nations have adopted targets for deep reductions in greenhouse gas emissions by certain dates, often 2050. However, these targets have not generally been based on realistic and detailed analysis of the energy, infrastructure, and economic transformations required to meet them.[†] A multidisciplinary group of scientists recently attempted to model and analyze the specific changes in infrastructure, technology, and governance and the costs of decarbonizing California, the world's sixth largest economy and twelfth largest emitter of greenhouse gases. California's gross domestic product (GDP) and greenhouse gas emissions are comparable to those of Western Europe and Japan. California has passed legislation that sets a target of reducing greenhouse gas emissions 80% below the 1990 level by 2050, a level consistent with the IPCC emissions reductions trajectory that would stabilize atmospheric greenhouse gas concentrations at 450 parts per million of CO_2 equivalent (CO_2-eq). This was the 2007 IPCC target for avoiding dangerous interference with the climate.[‡]

This analysis did not assume any lifestyle changes (like vegetarianism or bicycle transportation), which the researchers acknowledged could have a substantial effect on mitigation requirements and costs. Meeting the 80% emissions reduction would require nearly total decarbonization of electricity generation (replacing fossil fuels or removing the carbon from them), replacing most direct petroleum fuel uses with decarbonized electricity, and improving energy efficiency to levels and at rates never before achieved. Achieving these changes "will depend substantially on technologies that are not yet commercialized."[§] Decarbonizing the electricity grid is projected to require the use of nuclear power, renewable energy (mostly wind and solar), and CCS. The largest anticipated greenhouse gas reductions from expanded

---

[*] Johannes Bollen, Sebastiaan Hers, and Bob van der Zwaan, "An integrated assessment of climate change, air pollution, and energy security policy," *Energy Policy* 38, no. 8 (2010).
[†] Williams et al., "The technology path to deep greenhouse gas emissions cuts by 2050."
[‡] Ibid.
[§] Williams et al., "The technology path to deep greenhouse gas emissions cuts by 2050." 53.

electrification are expected in transportation, with 70% of miles traveled to be powered by decarbonized electricity, 20% by biofuels, and 10% by fossil fuels. Space heating, water heating, and industrial processes are also assumed to be fueled almost entirely with decarbonized electricity. None of the scenarios conceived could achieve a greenhouse gas reduction of more than 50% without the near total replacement of fossil fuels in transportation and other sectors with decarbonized electricity. Assuming even the most optimistic breakthroughs in renewable energy development, no more than 74% of California's energy needs could be met by wind, solar, and other renewables, even with unprecedented improvements in energy efficiency averaging 1.3% per year for forty years.[*]

Not all transportation can be electrified, so biofuels are projected to play a small but essential role in decarbonizing industrial economies. This will require major improvements in cellulosic and algal biofuel technology. Other technological breakthroughs required for this scenario to work include the commercial availability of CSS, on-grid energy storage, improved electric vehicle batteries, smart charging (timing battery charging to times of low grid demand), and improved appliances, buildings, cement manufacturing, electrical industrial boilers, and capture of emissions from industry (beyond CCS). The authors also cite the need for improved agriculture and forestry practices.[†]

The estimated cost of this infrastructure and technology transformation is $1.4 trillion from 2010 to 2050 and $1,200 per capita per year in 2050. The State of California is considered by the research team to be generally representative of developed economies in that they will share "the need to virtually eliminate fossil fuel in electricity supply and in final consumption, especially in vehicles and buildings" in order to achieve deep greenhouse gas reduction targets.[‡] The researchers also noted that switching fuel to electricity before the electricity grid is decarbonized would negate the emissions benefits of electrification. All of these technologies will need to be commercially ready and deployed in proper sequence in order to achieve the desired emissions reductions. "The logical sequence of deployment" is improved energy efficiency first, followed by decarbonization of electricity generation, and finally, replacing nearly all oil and gas with decarbonized electricity.[§]

If the energy efficiency of vehicles, buildings, appliances, consumer electronics, and industrial processes can be improved at rates never before achieved, *if* carbon-free renewable energy systems can be developed and deployed in the most optimistically projected time frames, *if* the electric grid can be decarbonized using technologies like CCS that are not yet available, *if* cellulosic biofuels can become commercially viable, *if* the resources

---

[*] Williams et al., "The technology path to deep greenhouse gas emissions cuts by 2050," 10.
[†] Williams et al., "The technology path to deep greenhouse gas emissions cuts by 2050," 53.
[‡] Williams et al., "The technology path to deep greenhouse gas emissions cuts by 2050," 55.
[§] Williams et al., "The technology path to deep greenhouse gas emissions cuts by 2050," 58.

are available to replace or retrofit nearly everything that runs on oil or gas with electric alternatives, *and if* this transformation takes place in every nation and every city on Earth, then in forty years greenhouse gas emissions may be reduced by 80% below the 1990 level. Remember also that this level of energy system transformation is projected to stabilize atmospheric concentrations of greenhouse gases at 450 parts per million, a level nearly 30% higher than the 350 parts per million identified by Hansen and many other climate scientists as the maximum concentration that will avoid catastrophic climate disturbance.[*]

The assumption that the required energy system transformation can or should proceed without lifestyle changes, even on the part of those living in opulence while at least a billion people live in absolute poverty, raises moral and practical questions of equity and absolute scarcity of resources. The richest 1% of adults owns 40% of global wealth and the wealthiest 10% own 85%, while the bottom half of the population owns about 1% of total global wealth.[†] Current levels of wealth and resource inequality cannot be maintained in a nongrowing economy with increasingly scarce energy and material resources without increasing conflict between the poor and the privileged.[‡] Clear signs of increasingly organized resistance and revolt by the poor in order to secure a livelihood, often centered on access to energy resources, are emerging globally.[§] It is likely that a redistribution of wealth and resources from the industrial north to the global south and from wealthy elites to the poor will be necessary if finite resources are to provide an adequate livelihood for 9 billion people. Growing the global economy to meet growing energy and material needs, although still the official international objective,[¶] is not physically possible in the long run,[**] and is increasingly problematic now.[††] It is not physically possible if economic growth continues to result in increased throughput of energy and material resources and in ever increasing levels of waste dumped into the ecosphere. I return to this dilemma below.

5.2.4 The Limitations of International Regulatory Frameworks

The 2011 UN Framework Convention on Climate Change (FCCC) meeting in Durban, South Africa, examined evidence that confirmed that global agriculture must produce more food to feed a growing world population. The growing threat to agricultural yields and food security posed by climate change,

---

[*] Hansen, *Storms of My Grandchildren*.
[†] Richard Heinberg, *The End of Growth: Adapting to Our New Economic Reality* (Gabriola, BC, Canada: New Society Publishers, 2011).
[‡] Heinberg, *The End of Growth*.
[§] Abramsky, *Sparking a Worldwide Energy Revolution*.
[¶] Food and Agriculture Organisation, *The State of the World's Land and Water Resources*.
[**] Daly, *Beyond Growth*.
[††] Heinberg, *The End of Growth*.

especially increased frequency of extreme droughts and flooding, has been confirmed by scientific assessments of global food production capacity and brought into stark relief by recent droughts and floods in Russia, Australia, Pakistan, and the Horn of Africa during the years leading up to the Durban meeting.[*] The global hunger now caused by poverty, food waste, and distribution problems is evolving into the threat of global hunger due to absolute food shortages caused by soil and water resource scarcities exacerbated by climate change.[†]

Alternative agricultural practices, like Farmer Managed Natural Regeneration (FMNR) and other agroforestry practices in Niger that have now benefitted more than 1.25 million households and increased grain production by 500,000 metric tons per year, or alternative agriculture efforts in Denmark that have reduced greenhouse gas emissions from agriculture while increasing productivity, were identified by the FCCC as areas of promise. These proven examples can be scaled up to address the need for both agricultural adaptations to climate change and mitigation of greenhouse gases. Although no formal agreement on agricultural policy was reached in Durban in 2011, progress was made in defining the terms "climate-smart agriculture" and "sustainable intensification." Several elements were identified as essential to an agriculture that contributes to both adaptation and mitigation. To be climate smart and contribute to sustainable intensification, agricultural systems must maintain or increase food, fiber, and fuel production; support livelihoods and build prosperity; sustain environmental resources and ecosystems; adapt to existing and future climate change; and sequester carbon and/or reduce greenhouse gas emissions.[‡] This is an important official recognition that fundamental changes to agricultural practices are needed globally to further the transition from practices that degrade ecosystems and emit greenhouse gases to ones that restore ecosystems and sequester carbon.

The 2011 Durban climate meeting also recognized a need for more research and funding for incentives to steer public and private investment into sustainable agricultural practices. The Adaptation Fund of the Kyoto Protocol, the Green Climate Fund, which is to provide $100 billion per year for adaptation and mitigation of climate change in developing countries, and the Clean Development Mechanism of the FCCC are all potential funding mechanisms for investment in improved agricultural practices.[§] Many regional, national, state, and local funding sources, both public and private, are also in place. These are important resources for investment in local implementation of improved land management practices and the development of no- or low-carbon biomass energy sources.

---

[*] J. R. Beddington et al., "What next for agriculture after Durban?," *Science* 335, no. 6066 (2012).
[†] Food and Agriculture Organisation, *The State of the World's Land and Water Resources*.
[‡] Beddington et al., "What next for agriculture after Durban?."
[§] Ibid.

Resources for implementing proven restorative agricultural practices are in danger of being diverted to a new "Green Revolution" devoted primarily to importing industrial agricultural technology to Africa, India, and South America. The original Green Revolution was widely credited with helping more than 150 million people out of poverty in developing nations from about 1970 to 1990 by increasing food production using fossil-based fertilizers, pesticides, and improved seed varieties. The actual record does not support the legend. These practices were accompanied by increased deforestation and large-scale displacement of peasants, especially in South America. Decreases in poverty and hunger were concentrated in China, with worsening conditions elsewhere. The ecological impacts of the first Green Revolution were decidedly negative and the poverty and food security outcomes were mixed, at best.[*]

Contemporary versions of Green Revolution projects have incorporated an important lesson from earlier mistakes and are directing resources to small landholders and women, avoiding the displacement of peasants that accompanied the first Green Revolution. A central component of the new projects, however, is the transfer of industrial agricultural technology, especially biotechnology, to farmers in developing nations. About 30% of the $1.3 billion in agricultural development grants made by the Bill and Melinda Gates Foundation (as of 2009) were for science and technology. Much of that was for promoting and developing seed biotechnologies. Monsanto, a leading developer of biotech products, is involved in many of these Gates-funded technology transfer projects.[†] Multinational corporations who profit from expanding global markets and well-intentioned philanthropists who rely heavily on industrial technology to address social problems may be able to garner more financial resources than those who support the spread of local knowledge and practices.

In contrast to the large investments required to make energy technology and infrastructure changes or to import industrial agricultural technologies, restorative agricultural changes are often less expensive than the practices they replace. Regeneration of native tree species and using the biomass produced to enhance soil carbon is less expensive than establishing a centralized tree nursery, planting imported trees, and buying commercial fertilizer. Planting legumes in rotation with grain crops is less expensive than purchasing nitrogen fertilizer. Restoring grassland with diverse mixtures of native perennial grassland species, including legumes, is less expensive than planting nonnative cool season grasses that require extensive ongoing fertilizer inputs. Investments in such practices also quickly produce tangible soil and biodiversity improvements.

---

[*] Raj Patel, Eric Holt-Giminez, and Annie Shattuck, "Ending Africa's hunger," *The Nation* 289, no. 8, September 21, 2009, 17–22.

[†] Patel et al., "Ending Africa's hunger," 17–22.

Many observers of the UN FCCC process, me included, have been frustrated by the series of failures—from Rio to Copenhagen and Durban—to produce binding agreements for reductions in greenhouse gas emissions or other important actions on climate change. Agreements by heads of state to reduce greenhouse gas emissions, if and when such agreements are achieved, should not be mistaken for taking concrete steps to make the transition to a no-carbon energy system or to land management practices that build soil and restore ecological diversity. The regulatory schemes to address carbon emissions developed thus far have accomplished little more than an illusion of change and a means of making money for a few.[*] The world's most fully developed and widely touted trading scheme for carbon, the European Union Emissions Trading Scheme (EU ETS), has "failed to reduce structural dependence on fossil fuels in Europe." If this seems to set the bar too high, the EU ETS has also failed "to spur research and development on low-carbon alternatives....It has not even slowed the growth of Europe's emissions."[†]

The Kyoto Protocol went into force in 2005 and was eventually ratified or agreed to by 190 nations. The agreement was intended to slow global warming by mandating a reduction of CO_2 and other greenhouse gasses among all industrialized countries by 5% below 1990 levels by 2012. By 2007 the collective emissions from industrial nations had dropped by 4% below 1990 levels, but eighteen of the forty industrial nations had higher emissions in 2009 than they did in 1990.[‡] The United States, the second largest emitter behind China (which was not included in the Kyoto emissions agreement), had increased its greenhouse gas emissions by 17% over 1990 levels by 2009. Spain, Australia, Canada, and several other nations had even higher percentage increases.[§] In 2009 the United States was still the lone holdout among the 190 nations that negotiated the accord. This was probably the central reason that the Kyoto Protocol "fell short of its primary aim—catalyzing serious emissions reductions by all major industrial powers."[¶] Global emissions of greenhouse gases continue to increase and, most importantly, they continue to accumulate in the atmosphere at a rate that greatly exceeds the capacity of terrestrial and marine ecosystems to absorb them. International regulatory schemes have thus far failed to provide evidence that they can be the catalyst for emissions reductions, much less the structural transformation of industrial societies, or even their energy systems.

The current climate and ecological situation requires that we rapidly reduce and then virtually eliminate fossil fuel use. Decreased overall use of energy and creating decentralized forms of energy from local biomass

---

[*] Lohmann, "Capital and climate change."
[†] Lohmann, "Capital and climate change," 655. Lohmann provides multiple citations for each claim.
[‡] Janet Raloff, "Climate might be right for a deal: Copenhagen negotiations will take steps toward a climate-stabilizing treaty," *Science News* 176, no. 12 (2009).
[§] Raloff, "Climate might be right for a deal."
[¶] Raloff, "Climate might be right for a deal, 17.

plants, geothermal, solar, wind, hydro, and other nonfossil sources that can be developed and operated under democratic local control and not necessarily tied to large grid systems are possible alternate steps toward decarbonizing the economy. This path does not require global agreement among parties with competing interests, including the powerful self-interest of those who profit from a centralized, fossil-fueled, grid-connected energy system. Carbon trading schemes create a new commodity (carbon) that serves as a new source of profit for the banking sector and others, primarily in rich, industrial societies positioned to benefit from yet another form of speculative finance.[*]

The UN FCCC, the UN Convention on Biodiversity, and other UN bodies provide critically important forums for international dialogue and cooperation. They serve to gather and disseminate indispensible collections of scientific and policy-relevant information. This is a vital role. Perhaps they will progressively become the locus of more concrete policies and actions on climate change and ecological restoration. However, international regulatory frameworks have thus far failed to make significant inroads into changing the structural conditions that continue to produce increasing carbon emissions and increasing atmospheric concentrations of greenhouse gases. The regulatory bodies of the UN are associations of the very same governments that view perpetual economic growth as an essential mandate. Expecting national governments committed to growth to take concrete steps through international agreements toward structural changes required to mitigate climate change and ecological degradation has proven to be slow, at best.

5.2.5 Local Action and a New Relationship with Nature

There are local movements on every continent that recognize and are operating on the understanding that climate solutions will not be found solely at the level of international regulatory frameworks and that the needed changes will reach far wider than energy systems. The dominant, official approaches to mitigating and adapting to climate change continue to focus on promoting regulation, while maintaining the basic structure of industrial capitalism, rather than on more fundamental changes in political and social relations.

> The kind of massive and rapid reductions in CO_2 emissions required (and the corresponding changes in energy production and consumption that are necessary for this to occur) will not be possible without extensive changes in production and consumption relations at a more general level, involving fundamental change in how humans interact with nature....The process of building a new energy system, based around a greatly expanded use of renewable energies, has the potential to make an important contribution to the construction of new relations of

---

[*] Lohmann, "Capital and climate change," Vandana Shiva, *Soil not Oil: Environmental Justice in a Time of Climate Crisis* (Cambridge, MA: South End Press, 2008).

production, exchange and livelihood that are based on solidarity, diversity, and autonomy, and are substantially more democratic and egalitarian than the current relations.[*]

The history of energy systems in industrial societies demonstrates clearly that energy-efficient technologies and renewable energy are not sufficient to alter the social relations of exploitation and the hierarchy of capitalism. Capitalist economies emerged in an era of windmills, water-powered mills, and sailboats.[†] Fossil fuels and seemingly "cheap" energy later made the rapid global expansion of industrial capitalism possible. Any structural transformation of industrial capitalism will require more than a transition away from fossil fuels. In describing some of the diverse local movements that are currently working to make a transition to postcapitalist forms of democratic and egalitarian communities, Kolya Abramsky harbors no illusions about the resistance to be encountered:

> States and corporations will do anything in their power to maintain capitalist social relations as *the* fundamental form of reproducing our livelihoods. Furthermore, the experience of capitalist renewable energy regimes of the past stands as a reminder that social relations of production, based on enclosures and exploitation, are not exclusively associated with fossil fuels and nuclear energy. There is nothing automatically emancipatory about renewable energies.[‡]

The "enclosures" mentioned by Abramsky signify the central importance of land, and its use and control, to the maintenance of capitalist social relations of exploitation. As Karl Polanyi noted in *The Great Transformation*, exploitation of both labor and land was a central and essential feature of the colonization of the world by the colonizers who later created industrial capitalism.

> Whether the colonist needs land for the sake of the wealth buried in it, or whether he merely wishes to constrain the native to produce a surplus of food and raw materials, is often irrelevant; nor does it make much difference whether the native works under some form of direct supervision of the colonist or only under some form of indirect compulsion, for in every and any case the social and cultural system of native life must be first shattered.[§]

Polanyi observed that the "subordination of the surface of the planet to the needs of an industrial society" required separating people from the soil,

---

[*] Abramsky, *Sparking a Worldwide Energy Revolution*, 7.

[†] Abramsky, *Sparking a Worldwide Energy Revolution*; David E. Nye, *Consuming Power: A Social History of American Energies* (Cambridge, MA: MIT Press, 1998).

[‡] Abramsky, *Sparking a Worldwide Energy Revolution*; Nye, *Consuming Power: A Social History of American Energies*, 13.

[§] Karl Polanyi, *The Great Transformation* (Boston: Beacon Press, 1957), 178.

dissolution of autonomous local economies, and the elimination of lands held in common, accomplished by a series of legislative acts, including the Enclosure Act of 1801 and its successors.* For our purposes, the importance of this history is to recognize that the private commodification of land and the control of the people who worked it were essential to the revolutionary transformation from feudal, precapitalist social forms to industrial, capitalist economies. The importance of these factors has not diminished. A new transition to postcapitalist economies and societies and transcending the growth imperative will also require a fundamental transformation in the relationship between humans and land. The enclosure process described by Polanyi continues today. The acquisition of land and natural resources in Africa, South America, and elsewhere by China, India, and other emerging and rapidly growing industrial economies is a prime example. Separating local people from the soil, the enclosure of lands held in common, and undermining autonomous local economies by expropriating scarce resources in foreign lands is now required by growing economies as they exceed the resource and ecological limits of their homeland.

It is not necessary to eliminate private ownership of land, water rights, and other resources to make changes in land use and land management practices. Access to land through ownership by those who want to restore ecosystems and who have the needed resources can be expanded. Collective democratic control over public lands to protect and preserve ecosystems and natural resources can be enhanced through existing institutional forms. This is one of the central political struggles being engaged in by those working to create democratic communities and no-carbon energy systems.† Legal and regulatory controls of ecologically destructive practices by individuals and corporations on private land also exist and can also be improved and expanded. These controls are national, state, and local and do not require complex global agreements. Their creation and improvement do require political participation and, as noted by Abramsky, will meet resistance by those benefiting from the current concentration of wealth and resources.

Government policies that remove perverse incentives and subsidies for destructive practices like mining and burning fossil fuels or promote soil degradation are an essential component of furthering and accelerating the transformation that has begun. Action is under way in nations, states, and local communities to replace perverse subsidies for ecologically destructive practices with new incentives for ecosystem protection and restoration, energy conservation and efficiency, public transportation, carbon-neutral or carbon-negative energy sources, and the expansion of publicly funded scientific research and education.‡

---

* Polanyi, *The Great Transformation*, 179–180.
† Abramsky, *Sparking a Worldwide Energy Revolution*.
‡ Stephen Henry Schneider, *Climate Change Science and Policy* (Washington, DC: Island Press, 2010).

A key question, not immediately answerable, is are we capable of acting rapidly enough to avoid catastrophic climate change and ecological collapse? Even if the answer is unknown, the evidence is clear that pursuing the needed energy transformation and implementing restorative land practices can minimize the damage and improve the likelihood of successful adaptation. Making changes now and as rapidly and extensively as possible does not guarantee avoidance of collapse, but organized communities implementing restorative land management practices, reducing their energy use, and developing nonfossil energy sources will be better prepared to adapt to climate change and resource scarcity than those who remain dependent on multinational corporations, or others, to meet their basic needs. A central aspect of the transformation required to mitigate and adapt to our changed circumstances involves moving beyond an abstract declaration that a new relationship between humans and nature is required. That new relationship must take concrete form in specific actions that build soil and sequester carbon, increase ecological diversity, decrease overall energy use, and create and use alternatives to fossil fuel. The practices discussed in Chapter 4 are currently existing examples that accomplish these goals. All can be replicated and adapted to local ecosystems.

Taking such actions does not require that everyone become a farmer or even a gardener. Those who do not have access to land or choose not to work directly with soil and plants can cultivate cooperative relationships with those who do. Enhancing general collective competence and productive capacity at the community level can make producing food, fiber, wood products, energy, and a host of other necessities a concrete part of local life instead of relying on commodities purchased from anonymous sources. Actively working to protect local sources of clean water, protecting and enhancing soil, forests, grasslands, wetlands, and all of the ecosystems upon which we depend takes the abstract passivity out of the idea of a "new relationship with nature."

There are diverse needs in each community and region for people with skills and talent for technology development, artistic expression and crafts, healthcare and wellness services, education, spiritual pursuits, construction and architecture, and production of food, fiber, wood, and energy—in short, the full range of human needs. The "easy" labor-saving lifestyle of globalized capitalism has proven to not be so easy or satisfying, especially for the unemployed, but even among those who achieve material success. Americans, for example, have higher incomes, larger houses, and more cars than they did in the 1950s, but the percentage of American describing themselves as happy peaked in the 1950s.*

Can we collectively implement the proven land use and management practices that sequester carbon, build soil, and enhance ecological diversity on a large enough scale and quickly enough to preserve viable forests,

---

* Heinberg, *The End of Growth.*

grasslands, and wetlands and to ensure food security for 9 billion people in 2050? Like the uncertainties in climate models regarding the future of terrestrial carbon sinks and sources, the outcome depends on human action from this time forward. There is time to act, not to preserve the world as it is, but to improve on what would likely occur in the absence of concerted action to enhance biosequestration and restore ecosystems, reduce energy use, and progressively eliminate fossil fuels. This is the best the Anthropocene has to offer. There are diverse opportunities for meaningful action. Business as usual is the most dangerous option. It has gotten us to the point where we now face "the perfect storm of ecological and social problems."[*]

5.2.6 The Growth Dilemma and Globalization

The growth imperative in industrial capitalist societies presents perhaps the most pervasive and difficult policy dilemma and barrier to transforming economies and societies to more sustainable forms. The dilemma is made up of two contradictory realities; unending growth of energy and material resource consumption is simply not possible in a finite world, yet the governments of essentially all nations have taken on the role of protecting and stimulating economic growth. Economic growth is understandably viewed as the means to reduce unemployment, meet the material needs of growing populations, and improve the living standards of the poor. The promotion of economic growth is bipartisan and pursued in poor developing nations as well as wealthy industrialized countries.

When confronted with the ecological damage and inevitable resource scarcities associated with expanding the throughput of material resources and wastes, which have universally accompanied economic growth,[†] the dominant response by economists has been to tout improved efficiency and to attempt to "decouple" material throughput and wastes from economic growth. Decoupling involves reconfiguring production and redesigning goods and services so that economic output is less dependent on material throughput. This is a powerful idea that has had some success in relative terms. The energy intensity of the global economy, the energy consumed per unit of economic output, has declined by 33% since 1970.[‡] This global average reflects improvements primarily in the United States, Western Europe, and the United Kingdom and masks great variability in energy intensity elsewhere. Energy intensity has increased during the past twenty-five years in southern Europe and the Middle East. China's energy intensity has begun to rise in recent years after falling by 70% in the decades prior to 2000.[§]

---

[*] Vidal, "Civilisation faces 'perfect storm of ecological and social problems.'"
[†] Daly, *Beyond Growth*.
[‡] IPCC (2007), as cited by Jackson, *Prosperity Without Growth*.
[§] *Prosperity Without Growth*.

The overall and uneven decline in energy intensity over the last several decades has been accompanied by similar reductions in material resource intensity and carbon intensity (emissions per unit of production). Despite declining energy, material, and carbon intensities in the global economy, however, CO_2 emissions from burning fossil fuels have increased by 80% since 1970. Emissions in 2010 were 40% higher than they were in 1990.[*] *Global economic growth has overwhelmed improvements in energy and material efficiency.*

Absolute decoupling, a net reduction in total energy and material throughput, has not occurred. The history of economic growth and increased material throughput "provides little support for the plausibility of decoupling as a sufficient solution to the dilemma of growth."[†] Substituting new materials for those that have been depleted, another common response by economists, is no more plausible as a complete answer. There is no substitute for soil[‡] or other essential and finite natural resources (natural capitol), especially healthy terrestrial and marine ecosystems that act as a sink for the wastes associated with increased material throughput.

The "compulsive drive to grow" has long been recognized as an essential feature of capitalist economies, as has the "tension between capitalist growth and the environment."[§] Social theorists argued in the nineteenth and early twentieth centuries that the increasing consumption of natural resources in capitalist societies would eventually approach the limits of the finite resource base. Following World War II and intensifying after the collapse of the Soviet Union in 1989, a new global capitalism emerged that greatly accelerated and extended the global reach of economic growth. This "neoliberal globalization," characterized by a belief in a "free market" that is assumed by its advocates to be essentially "immune to ecological limits," led to a period of high consumption and high rates of global economic growth until the financial collapse and recession of 2008.[¶] This period of accelerated global economic growth, during the second half of the twentieth century, was the same time period identified by James Hansen as when human emissions finally overwhelmed natural climate variations and became the dominant influence on global climate.[**] The long life of CO_2 in the atmosphere and the inertia of the climate system, due primarily to the large mass of the oceans,[††] mean that we have not yet experienced the full climate and ecological effects of postwar globalization and the associated intensification of consumption and emissions.

---

[*] *Prosperity Without Growth.*
[†] IPCC (2007), as cited by Jackson, *Prosperity Without Growth*, 75.
[‡] Jackson, *Consulting the Genius of the Place.*
[§] Antonio, "Climate change, the resource crunch, and the global growth imperative," 4.
[¶] Antonio, "Climate change, the resource crunch, and the global growth imperative," 8.
[**] Hansen, *Storms of My Grandchildren.*
[††] Brian Dawson and Matt Spannagle, *The Complete Guide to Climate Change* (New York: Routledge, 2009).

In 1972 Donella Meadows et al. published the first edition of *Limits to Growth*, followed by *Beyond the Limits* in 1992 and *Limits to Growth: The 30-Year Update* in 2004.* These books quantitatively trace the progression of the extent to which the growth of human population and economies has led to "overshoot, an expansion in demands on the planet's sources and sinks above levels that can be sustained."† In 1972 the evidence supported a conclusion that growth could safely continue and the global population and economy were well below the Earth's carrying capacity. By 1992 the research team cited evidence that the Earth's carrying capacity had been exceeded by approximately 20%. In 2004 the human ecological footprint was still increasing and the "general awareness of this predicament" was "hopelessly limited." The authors concluded that it will take decades, at least, to make the changes needed to "bring the ecological footprint back below the long-term carrying capacity of the planet."‡

How can this be accomplished? How can the needs of a growing human population be met without further exacerbating the ecological degradation, resource scarcities, carbon cycle disruption, and climate change associated with economic growth? This is emerging as a central and perhaps one of the most difficult questions of our new epoch, the Anthropocene. There is, of course, no complete answer available. Some components of a strategy are in place. The intensification of agricultural production on existing cropland without further degradation of soil and water resources, as proposed by the FAO,§ is an essential, and challenging, part of any plan to meet future needs. Food production is one area where increased material output will be required as the population grows, even if the need for a 70% increase in production by 2050 estimated by the FAO can be reduced somewhat by changing diets and decreased inequality in access to food. Reducing meat consumption and the amount of grain fed to animals in industrial societies, for example, could potentially make more grain available for humans in developing nations.¶ Enhancing biosequestration of carbon on agricultural lands and increasing ecological diversity are required to meet the twin goals of higher agricultural production and reduced soil degradation. Restorative land management is an essential prerequisite for maintaining and increasing food production.

Relative decoupling—working to reduce the energy, material, and emissions density of economic production—has demonstrated the potential to

---

* Donella H. Meadows, *The Limits to Growth: A Report for the Club of Rome's Project on the Predicament of Mankind* (New York: Universe Books, 1972); Donella H. Meadows, Dennis L. Meadows, and Jørgen Randers, *Beyond the Limits: Confronting Global Collapse, Envisioning a Sustainable Future* (Post Mills, VT: Chelsea Green, 1992); Donella H. Meadows, Jørgen Randers, and Dennis L. Meadows, *Limits to Growth: The 30-Year Update* (White River Junction, VT: Chelsea Green, 2004).
† Meadows et al., *Limits to Growth: The 30-Year Update*, xii.
‡ Meadows et al., *Limits to Growth: The 30-Year Update*, xiv–xv.
§ Food and Agriculture Organisation, *The State of the World's Land and Water Resources*.
¶ Daly, *Beyond Growth*.

make a contribution to reducing throughput. Improved energy efficiency in transportation and housing, decarbonization of electricity, and replacing liquid petroleum fuels with electricity and biofuels are the focus of substantial research and development, and will undoubtedly be a part of the broad portfolio of needed changes. Substitution—replacing scarce, degraded, or damaging processes and materials with less scarce or less damaging alternatives—is not a panacea, but it can make a contribution. Displacing some fossil fuel–intensive materials, like plastic, steel, and concrete, with wood or other renewable, less energy- and emissions-dense products is an example. The partial displacement of coal with biomass energy is an example of a bridge strategy that can reduce the damage of existing technology as better alternatives are in development. An engineering movement to change the design of products—from automobiles to consumer electronics to industrial infrastructure—so that they can be repaired, recycled, and reused rather than discarded has begun and incorporates components of decoupling, substitution, and displacement.[*] Each of these changes is necessary and important, but also is likely insufficient to produce the needed end to the growth of energy, material, and waste throughput if economic growth, as currently defined, continues.[†]

5.2.7 A New Economics and the Politics of Denial

A net decrease in energy and material throughput and a net reduction of the wastes produced by global economic activity requires rethinking, redefining, and redirecting growth.[‡] Can "growth" be redirected toward an improvement in quality of life largely decoupled from increasing quantity of material throughput? There is guidance available from a strong and long-lived tradition in economic thought and practice that recognizes ecological limits and rejects the possibility of unending material growth.

The pioneering work of E. F. Schumacher placed the use of land at the center of his economics: "Study how a society uses its land, and you can come to pretty reliable conclusions as to what its future will be....The land carries the topsoil, and the topsoil carries an immense variety of living beings, including man."[§] Schumacher was citing the work of ecologists and addressing the "ecological problem" caused by population density, soil erosion, deforestation, the depletion of minerals, wildlife, and fish, and an emerging full world with "no new lands to move to" in 1973.[¶] This was about a quarter century

---

[*] William McDonough and Michael Braungart, *Cradle to Cradle: Remaking the Way We Make Things*, 1st ed. (New York: North Point Press, 2002).

[†] Heinberg, *The End of Growth*.

[‡] Heinberg, *The End of Growth*; Antonio, "Climate change, the resource crunch, and the global growth imperative."

[§] E. F. Schumacher, *Small Is Beautiful: Economics as if People Mattered* (New York: Harper & Row, 1973), 102.

[¶] Schumacher, *Small Is Beautiful*, 103.

into the post–World War II global economic boom. Building on Schumacher's work, John Michael Greer has recently argued that the United States could benefit from adopting the idea of "appropriate technology" that has been successful in helping many in the developing world. The idea of appropriate technology that grew from Schumacher's "intermediate technology" holds that "relatively simple technologies, powered by locally available energy sources and drawing on locally available raw materials, could provide paying jobs and an improved standard of living for working people."[*] Relying more on local resources and less on global markets, precisely the opposite of recent globalization trends, is a central requirement of creating economies that do not exceed ecological limits.[†]

The central ideas needed for designing economic systems oriented toward "qualitative improvement without quantitative increase" have been evolving since at least 1857 and the work of John Stewart Mill.[‡] Mill's concept of a "stationary state" economy, however, envisioned the end of population growth as a requirement for achieving a condition of zero growth in physical throughput.[§] A global population expected to grow from 7 billion in 2012 to 9 billion by 2050 complicates the growth dilemma. Increases in food production and more equitable access to food must be integrated with an overall reduction in material throughput and waste.

The general outlines of economic systems that reduce throughput of energy and material resources, and produce less waste—that do not exceed ecological limits—are clear and generally agreed upon by steady-state, ecological, no-growth, and postgrowth economists. Richard Heinberg, of the Post Carbon Institute, offers the following summary as a "best-case scenario":

> The economy of the future will necessarily be steady-state, not requiring constant growth. It will be based on the use of renewable resources harvested at a rate slower than that of natural replenishment; and on the use of non-renewable resources at declining rates, with metals and minerals recycled and re-used whenever possible. Human population will have to achieve a level that can be supported by resources used this way, and that level is likely to be significantly lower than the current one.[¶]

There are, of course, many details missing from this scenario. A population level that does not exceed the carrying capacity of the Earth and how that level is achieved stand out as recurring unanswered questions. Ecological restoration of soil carbon and functioning ecosystems will help maintain carrying capacity. Global population growth rates are declining and have

---

[*] Greer, *The Wealth of Nature*, 175.
[†] Greer, *The Wealth of Nature*, 175; Heinberg, *The End of Growth*; Jackson, *Prosperity Without Growth*.
[‡] Daly, *Beyond Growth*, 3.
[§] Daly, *Beyond Growth*, 3.
[¶] Heinberg, *The End of Growth*, 280–281.

been reduced in recent years without draconian measures. Providing access to education and economic resources to girls and women has a proven record of lowering birth rates.[*]

A vision of what a steady-state economy might look like is in place and has many advocates. The vision is not lacking. Changing the policies of governments and the practices of industry away from conventional throughput and waste-accelerating economic growth is now required. This and the financial and ideological interests reflected in the growth imperative are the overriding barriers to the structural transformation of industrial capitalist economies and societies. The market and "free trade" have failed to make the required changes and the ecological consequences are accelerating.[†]

Herman Daly, one of the founders of contemporary ecological or steady-state economics, points out that classical economists like Mill, Ricardo, and Malthus all recognized demographic and ecological limits related to the fixed amount of land on the planet. The recognition of physical limits to growth was effectively banished from mainstream neoclassical economics following the Industrial Revolution.[‡] Daly argues a "change in vision" involving the replacement of the "economic norm of quantitative expansion (growth) with that of qualitative improvement (development)" is required in order to move toward societies that live within ecological limits. He also notes that there "are enormous forces of denial aligned against this necessary shift in vision and analytic effort."[§] Since Daly published these prescient words in 1996 the "forces of denial" have expanded beyond the economists and policymakers wedded to growth to include a well-financed lobbying and legislative effort to deny the reality of anthropogenic climate change and the full range of ecological crises facing industrial society.[¶]

The organized deniers of anthropogenic climate change and ecological problems have had a significant effect on policy, especially in the United States. This political opposition to scientific evidence and governmental regulation is supported by well-resourced think tanks and funded by those who profit from the continuation of a global energy system powered by fossil fuels and a range of corporations and trade groups interested in avoiding environmental regulation and the associated costs. The Heartland Institute, for example, is a leading organizer of opposition to recognizing anthropogenic climate change in the United States, has a long history of promoting the interests of large corporations, and identifies its purpose as

---

[*] L. R. Brown, *World on the Edge: How to Prevent Environmental and Economic Collapse* (New York: W. W. Norton, 2011).

[†] Herman E. Daly and Joshua C. Farley, *Ecological Economics: Principles and Applications*, 2nd ed. (Washington, DC: Island Press, 2010).

[‡] Daly, *Beyond Growth*.

[§] Daly, *Beyond Growth*, 1.

[¶] Naomi Oreskes and Erik M. Conway, *Merchants of Doubt: How a Handful of Scientists Obscured the Truth on Issues from Tobacco Smoke to Global Warming*, 1st U.S. ed. (New York: Bloomsbury Press, 2010); James Lawrence Powell, *The Inquisition of Climate Science* (New York: Columbia University Press, 2011).

being "to discover, develop, and promote free-market solutions to social and economic problems."[*]

Those leading the organized opposition to recognizing anthropogenic climate change, and to environmental regulation generally, use many of the same tactics and even several of the same hired "experts" previously used by tobacco companies in their concerted, and for decades successful, efforts to avoid the regulation of tobacco. The same tactics and several of the same hired experts were also used by the manufacturers of chlorofluorocarbons to oppose regulations designed to address stratospheric ozone depletion, and by coal companies and utilities who denied that sulfur dioxide was the primary cause of acid rain. Although scientific evidence eventually prevailed in each of these cases, deadly delays have been a recurring result of the doubt instilled by those with a financial and ideological interest in continued use of hazardous products and ecologically damaging industrial practices.[†]

Those who defend industrial capitalism by denying the significance of climate change and related problems and by politicizing scientific evidence on these issues have also successfully polarized public opinion in the United States. Increasing percentages of the American population, especially those who identify themselves as conservatives or Republicans, have come to doubt the reality, causes, and serious dangers posed by anthropogenic climate change.[‡]

The scientific evidence that the Earth is warming and that human actions are almost certainly the primary cause obviously constitutes a perceived threat to business as usual—and rightfully so. The climate crisis and emerging ecological collapse have profound implications for the future of an industrial civilization organized for unending growth of material consumption fueled by ever larger quantities of fossil energy. The deniers and their corporate sponsors are correct in recognizing that the evidence accumulated by climate scientists over the past two decades has the potential to undermine the continued profitable extraction and use of fossil fuels and the myriad of industrial processes and products dependent on oil, gas, and coal. As long as the health, ecological, and economic costs of burning fossil fuels and industrial land management practices that degrade ecosystems are not fully and explicitly recognized in environmental policy or even economic accounting,[§] the deniers of climate science and the opposition to environmental regulation remain successful.

The U.S. Environmental Protection Agency (EPA), as the agency primarily responsible for enforcing the Clean Air Act and other environmental protection laws, has been a primary target of those opposed to environmental

---

[*] Powell, *The Inquisition of Climate Science*, 10.
[†] Oreskes and Conway, *Merchants of Doubt*.
[‡] Aaron M. McCright and Riley E. Dunlap, "The politicization of climate change and polarization in the American public's views of global warming, 2001–2010," *Sociological Quarterly* 52, no. 2 (2011).
[§] Greer, *The Wealth of Nature*.

regulation. There have been repeated efforts by Republicans in the U.S. Congress to undermine the EPA's authority to regulate greenhouse gases. This has continued despite a 2007 U.S. Supreme Court ruling, in *Massachusetts v. EPA*, confirming the EPA's legal duty to protect citizens from dangerous CO_2 pollution under the 1970 Clean Air Act.[*]

The EPA is also a favorite target of antiregulatory legislation introduced and passed in many state legislatures with the backing of the same corporate interests active at the federal level. In my home state of Kansas, the 2011 Kansas House of Representatives passed a resolution urging Congress "to prohibit the EPA from regulating greenhouse gas emissions" and to accomplish this "by any means necessary," including defunding the agency.[†] Exxon-Mobil, Kansas-based energy conglomerate Koch Industries, and oil and coal trade groups were prominent members of the Natural Resources Task Force of an organization called the American Legislative Exchange Council (ALEC) that drafted the "model legislation" from which the Kansas resolution was derived. ALEC drafts about 1,000 pieces of model legislation each year for introduction in state legislatures and, according to ALEC, about 200 of those become law.[‡]

ALEC, a nonprofit organization with corporate funding from Wal-Mart, Exxon-Mobil, Koch, AT&T, and others, has a mission of "promoting free markets" and "limited government." The Center for Media and Democracy considers ALEC a public policy front for corporate interests and is closely monitoring the bills ALEC has written for state legislatures. Common Cause has requested an IRS investigation into possibly revoking ALEC's tax-exempt status. In the meantime, ALEC's corporate members have contributed more than $317 million to state election campaigns during the past decade and continue to see their "model legislation" enacted into law.[§]

Sociologists Robert J. Antonio and Robert J. Brulle have provided an analysis of climate change denial and the partisan split over global warming in the U.S. Congress and among U.S. citizens that places current challenges to the EPA's right to regulate greenhouse gases and the more general opposition to action on climate change into historical and political context. These attacks are part of an ongoing "neoliberal" antienvironmentalism that has recently elevated climate change denial to the top of a broader agenda against environmental regulation.[¶] Neoliberalism is the contemporary version of American market liberalism, which has a policy orientation "stressing unfettered capitalism, strong property rights, and a minimal safety net,"

---

[*] Legal Information Institute, "Massachusetts v. EPA," http://www.law.cornell.edu/supct/html/05-1120.ZS.html

[†] Scott Rothschild, "ALEC has stong ties to Kansas legislature," *Lawrence Journal World*, July 25, 2011, 3A.

[‡] Rothschild, "ALEC has stong ties to Kansas legislature," 3A–5A.

[§] Rothschild, "ALEC has stong ties to Kansas legislature," 3A–5A.

[¶] Robert J. Antonio and Robert J. Brulle, "The unbearable lightness of politics: climate change denial and political polarization," *Sociological Quarterly* 52, no. 2 (2011).

as opposed to social liberalism, which supports "modest state intervention, redistribution, and welfare provision."[*]

Antonio and Brulle point out that the U.S. form of neoliberalism (commonly also referred to as conservatism) also has a transnational version, often embodied in the economic growth promotion and public-sector austerity (government budget cutting) policies of the World Bank and the International Monetary Fund. Climate change denial and discrediting climate science have become pivotal to the antiregulatory cause of neoliberals. They fear that if anthropogenic climate change and ecological decline are accepted as real by policymakers, it could become the rationale for wholesale government intervention and regulation. Antonio and Brulle conclude that an effective neoliberal campaign to instill a widespread fear of government regulation, a near universal and bipartisan agreement on the need for economic growth, and a recent emphasis on debt reduction make major policy action on climate change in the U.S. unlikely in the immediate future.[†]

The campaign to instill fear of government regulation appears to be reaching new levels in Kansas. The 2012 Kansas Legislature's House Committee on Energy and Utilities sponsored a Resolution (House Resolution 6018) "opposing and exposing the radical and destructive United Nations Agenda 21." The resolution identifies UN Agenda 21, which was created in 1992 at the UN Conference on Environment and Development in Rio de Janeiro, as "a comprehensive plan of extreme environmentalism, social engineering and global political control." This agenda "is being covertly pushed into local communities throughout the United States of America through the International Council of Local Environmental Initiatives through local 'sustainable development' policies such as Smart Growth, Wildlands Project, Resilient Cities, Regional Visioning Projects and other 'Green' or 'Alternative' projects." The resolution states that the UN's "plan of radical so-called 'sustainable development' views the American way of life of private property ownership, single family homes, private car ownership, individual travel choices and privately owned farms as destructive to the environment." There is also a warning of impending "socialist and communist redistribution of wealth" to accomplish the radical environmental goals associated with "sustainable development."[‡]

Kansas House Resolution 6018 (2012) is an extreme example of fear mongering and contains more than a hint of conspiracy theory intrigue. That it was introduced in the Kansas State Legislature, sponsored by the House Committee on Energy and Utilities, and afforded a hearing is, I think, one indication that views that would have been considered an outrageous expression of political paranoia (or theater) a few years ago have become part of mainstream political

---

[*] Antonio and Brulle, "The unbearable lightness of politics," 195.
[†] Antonio and Brulle, "The unbearable lightness of politics," 200.
[‡] Committee on Energy and Utilities, "House Resolution 6018," Kansas Legislature, Topeka, Kansas (2012).

culture in Kansas and much of the United States. This may reflect recognition that industrial society is under profound threat. Denial of the accumulating evidence of anthropogenic climate change and global ecological decline is one response to this existential threat. This is admittedly speculative and not offered as a definitive characterization of denial. The financial and ideological underpinnings of the organized denial of anthropogenic climate change as well as the tactics used for distorting and undermining climate and environmental science have been carefully documented.[*]

There are complex and powerful financial and ideological interests represented by the political opposition to policies aimed at reducing greenhouse gas emissions, energy system transformation, and a steady-state economy. These policies and environmental regulations are viewed, most fundamentally, as threats to economic growth. The "forces of denial" identified by Herman Daly are committed to an unfettered "free market" and unending economic growth.[†]

The Anthropocene epoch, the age of man, has been created by a profoundly dangerous human alteration of the Earth's biological, chemical and biophysical processes. Human population growth and economic expansion have exceeded the ecological carrying capacity of the planet. The dilemma of growth-finite ecosystems, degraded and depleted soil and water resources and overloaded atmospheric and ocean sinks in the context of population growth, and expanding throughput of fossil energy, material resources and waste—will define the coming decades. The human actions that have created this dilemma are being recorded in the geological record of industrial civilization. Will that record also show that we recognized in time that directing our energies to growth and to the balance sheet for next fiscal year is not more important and does not negate the need to also prepare for 2050 and beyond?

References

Abramsky, K. 2009. *Sparking a Worldwide Energy Revolution: Social Struggles in the Transition to a Post-Petrol World*. Oakland, CA: AK Press.

Antonio, R. J. 2009. "Climate change, the resource crunch, and the global growth imperative." In *Current Perspectives in Social Theory*, vol. 26, ed. Harry F. Dahms, 3–73. Bingley, UK: Emerald Group Publishing.

Antonio, R. J., and R. J. Brulle. 2011. "The unbearable lightness of politics: climate change denial and political polarization." *Sociological Quarterly* 52(2):195–202.

---

[*] Oreskes and Conway, *Merchants of Doubt*; Powell, *The Inquisition of Climate Science.*
[†] Daly, *Beyond Growth.*

Beddington, J. R., M. Asaduzzaman, M. E. Clark, A. Fernández Bremauntz, M. D. Guillou, D. J. B. Howlett, M. M. Jahn, et al. 2012. "What next for agriculture after Durban?" *Science* 335(6066):289–290.

Bollen, J., S. Hers, and B. van der Zwaan. 2010. "An integrated assessment of climate change, air pollution, and energy security policy." *Energy Policy* 38(8):4021–4030.

Brown, L. R. 2011. *World on the Edge: How to Prevent Environmental and Economic Collapse*. New York: W. W. Norton.

Committee on Energy and Utilities. 2012. "House Resolution 6018." Kansas Legislature, Topeka, Kansas.

Daly, H. E. 1996. *Beyond Growth: The Economics of Sustainable Development*. Boston: Beacon Press.

Daly, H. E., and J. C. Farley. 2010. *Ecological Economics: Principles and Applications*, 2nd ed. Washington, DC: Island Press.

Dawson, B., and M. Spannagle. 2009. *The Complete Guide to Climate Change*. New York: Routledge.

Faunce, W. A. 1981. *Problems of an Industrial Society*, 2nd ed. New York: McGraw-Hill.

Flannery, T. F. 2005. *The Weather Makers: How Man Is Changing the Climate and What It Means for Life on Earth*, 1st U.S. ed. New York: Atlantic Monthly Press.

———. 2009. *Now or Never: Why We Must Act Now to End Climate Change and Create a Sustainable Future*, 1st ed. New York: Atlantic Monthly Press.

Food and Agriculture Organisation. 2011. *The State of the World's Land and Water Resources for Food and Agriculture: Summary Report*. Rome: United Nations.

Greer, J. M. 2011. *The Wealth of Nature: Economics as if Survival Mattered*. Gabriola, BC, Canada: New Society Publishers.

Hansen, J. E. 2009. *Storms of My Grandchildren: The Truth About the Coming Climate Catastrophe and Our Last Chance to Save Humanity*, 1st U.S. ed. New York: Bloomsbury USA.

Heinberg, R. 2011. *The End of Growth: Adapting to Our New Economic Reality*. Gabriola, BC, Canada: New Society Publishers.

Jackson, T. 2011. *Prosperity Without Growth: Economics for a Finite Planet*. London: Earthscan.

Jackson, W. 2010. *Consulting the Genius of the Place: An Ecological Approach to a New Agriculture*. Berkeley, CA: Counterpoint Press.

Kansas Association of Conservation Districts. 2012. *Conservation in Kansas: An Overview*. Lawrence: Kansas Association of Conservation Districts.

Legal Information Institute. 2012. "Massachusetts v. EPA." http://www.law.cornell.edu/supct/html/05-1120.ZS.html

Lohmann, L. 2011. "Capital and climate change." *Development and Change* 42(2):649–668.

McCright, A. M., and R. E. Dunlap. 2011. "The politicization of climate change and polarization in the American public's views of global warming, 2001–2010." *Sociological Quarterly* 52(2):155–194.

McDonough, W., and M. Braungart. 2002. *Cradle to Cradle: Remaking the Way We Make Things*, 1st ed. New York: North Point Press.

Meadows, D. H. 1972. *The Limits to Growth: A Report for the Club of Rome's Project on the Predicament of Mankind*. New York: Universe Books.

Meadows, D. H., D. L. Meadows, and J. Randers. 1992. *Beyond the Limits: Confronting Global Collapse, Envisioning a Sustainable Future*. Post Mills, VT: Chelsea Green.

Meadows, D. H., J. Randers, and D. L. Meadows. 2004. *Limits to Growth: The 30-Year Update*. White River Junction.

National Research Council. 2010. *Informing an Effective Response to Climate Change: America's Climate Choices*, ed. Board on Atmospheric Sciences and Climate. Washington, DC: National Academies Press.

Nelson, E., S. Polasky, D. J. Lewis, A. J. Plantinga, E. Lonsdorf, D. White, D. Bael, and J. Lawler. 2008. "Efficiency of incentives to jointly increase carbon sequestration and species conservation on a landscape." *Proceedings of the National Academy of Sciences of the United States of America* 105(28):9471–9476.

Novak, K. 2012. Personal communication. Feb. 10, 2012.

Nye, D. E. 1998. *Consuming Power: A Social History of American Energies*. Cambridge, MA: MIT Press.

Ontario Ministry of Natural Resources. 2012. "Ontario Ministry of Natural Resources." http://www.mnr.gov.on.ca/en/

Oreskes, N., and E. M. Conway. 2010. *Merchants of Doubt: How a Handful of Scientists Obscured the Truth on Issues from Tobacco Smoke to Global Warming*, 1st U.S. ed. New York: Bloomsbury Press.

Pan, Y., R. A. Birdsey, J. Fang, R. Houghton, P. E. Kauppi, W. A. Kurz, O. L. Phillips, et al. 2011. "A large and persistent carbon sink in the world's forests." *Science* 333(6045):988–993.

Patel, R., E. Holt-Giminez, and A. Shattuck. 2009. "Ending Africa's hunger." *The Nation* 289(8):17–22.

Polanyi, K. 1957. *The Great Transformation*. Boston: Beacon Press.

Post, W. M, J. Amonette, R. A. Birdsey, C. T. Garten, Jr., R. L. Graham, Dr. R. C. Izaurralde, P. M. Jardine, et al. 2009. "Terrestrial carbon sequestration: science for enhancement and implementation." In *Carbon Sequestration and Its Role in the Global Carbon Cycle*, ed. B. J. McPherson and E. T. Sundquist. Washington, DC: American Geophysical Union.

Powell, J. L. 2011. *The Inquisition of Climate Science*. New York: Columbia University Press.

Raloff, J. 2009. "Climate might be right for a deal: Copenhagen negotiations will take steps toward a climate-stabilizing treaty." *Science News* 176(12):16–20.

Rothschild, S. 2011. "ALEC has stong ties to Kansas legislature." *Lawrence Journal World*, July 25, 2011, 3A–5A.

Running, S. W., R. R. Nemani, J. R. G. Townshend, and D. D. Baldocchi. 2009. "Next-generation terrestrial carbon monitoring." In *Carbon Sequestration and Its Role in the Global Carbon Cycle*, ed. B. J. McPherson and E. T. Sundquist. Washington, DC: American Geophysical Union.

Schneider, S. H. 2010. *Climate Change Science and Policy*. Washington, DC: Island Press.

Schumacher, E. F. 1973. *Small Is Beautiful: Economics as if People Mattered*. New York: Harper & Row.

Shiva, V. 2008. *Soil Not Oil: Environmental Justice in a Time of Climate Crisis*. Cambridge, MA: South End Press.

Tilman, D., J. Hill, and C. Lehman. 2006. "Carbon-negative biofuels from low-input high-diversity grassland biomass." *Science* 314(5805):1598–1600.

U.S. Department of Agriculture. 2011. "Conservation Reserve Program: Status-End of December 2011." Edited by Farm Service Agency, 2011.

———. 2011. "FY 2011. Budget Summary and Annual Performance Plan." http://www.obpa.usda.gov/budsum/FY11budsum.pdf

———. "Conservation Programs." http://www.fsa.usda.gov/conservation

———. "Environmental Quality Incentives Programs." http://www.nrcs.usda.gov/wps/portal/nrcs/main/national/programs/financial/eqip

————. "NRCS Conservation Programs," http://www.nrcs.usda.gov/programs/

Vidal, J. 2012. "Civilisation faces 'perfect storm of ecological and social problems.'" *The Guardian*, February 20, 2012, 1–2, http://www.guardian.co.uk/environment/2012/feb/20/climate-change-overconsumption

Williams, J.H., A. DeBenedictis, R. Ghanadan, A. Mahone, J. Moore, W. R. Morrow, S. Price, and M. S. Torn. 2012. "The technology path to deep greenhouse gas emissions cuts by 2050: the pivotal role of electricity." *Science* 335(6064):53–59.

Appendix A: Measures and Conversion Units

1 metric ton (t) = 1,000 kilograms (kg) = 2,205 pounds (lb) = 1.1 U.S. ton
1 gigaton (Gt) = 1,000,000,000 (t) (1 billion metric tons) = .1 petagram (Pg)
1 hectare (ha) = 2.471 acres = 10,000 square meters (m^2)
1 square kilometer (km^2) = 100 hectares (ha) = 0.386 square miles (mi^2)
1 metric ton carbon = 3.67 metric tons carbon dioxide
1 metric ton carbon dioxide = 0.273 metric ton carbon
1 metric ton per hectare = 892 pounds per acre
1 meter (m) = 39.37 inches
1 centimeter (cm) = 0.39 inches
1 kilometer (km) = 0.62 miles
1 gallon (gal) = 3.785 liters (L)
1 inch = 2.540 centimeters (cm)

Sources

Intergovernmental Panel for Climate Change, "Land Use, Land-Use Change and Forestry," 2007, http//www.ipcc.ch/ipccreports/sres/land_use/index.php?idp=12

Webster's Third International Dictionary and Seven Language Dictionary, vol. 2, ed. Merriam Webster (Chicago: Encyclopedia Britannica, Inc., 1976), 1424.

Appendix B: Surface Albedo

Black road	0.05–0.10
Coniferous forest	0.05–0.15
Deciduous forest	0.10–0.20
Cities	0.14–0.18
Field crops	0.15–0.25
Grassland	0.16–0.20
Desert	0.25–0.30
Sea ice	0.30–0.40
Clouds	0.04–0.90
Old snow	0.40–0.70
Fresh snow	0.75–0.95

Albedo is the measure of the proportion of sunlight that is reflected from a surface. The numeric values represent the percentage of light reflected on a scale from zero to one, where zero represents total absorption and one represents total reflection. (The scale is not exact; a surface with an albedo of exactly zero would be invisible.)

The average albedo for the Earth is about 0.30 (30%). The values given are typical for various surfaces and can vary widely. Color, texture, and the angle at which the sun hits a surface determine its albedo. Lighter surfaces are more reflective and absorb less radiation (and heat) than darker surfaces.

Sources

Allaby, M. 2005. *A Dictionary of Ecology*, 3rd ed. Oxford: Oxford University Press, 13.
Dawson, B., and M. Spannagle. 2009. *The Complete Guide to Climate Change*. New York: Routledge, 14–15.

Index